Project
Puffin

Project Puffin

*The Improbable Quest to Bring a
Beloved Seabird Back to Egg Rock*

STEPHEN W. KRESS & DERRICK Z. JACKSON

Yale UNIVERSITY PRESS

New Haven and London

Yale University Press books may be purchased in quantity for
educational, business, or promotional use. For information, please
e-mail sales.press@yale.edu (US office) or sales@yaleup.co.uk
(UK office).

Designed by Sonia Shannon.
Set in Minion type by Integrated Publishing Solutions,
Grand Rapids, Michigan.
Printed in the United States of America.

Library of Congress Cataloging-in-Publication Data
Kress, Stephen W.
Project puffin : the improbable quest to bring a beloved seabird
back to Egg Rock / Stephen W. Kress and Derrick Z. Jackson.
pages cm
Includes bibliographical references and index.
ISBN 978-0-300-20481-0 (clothbound : alk. paper)
1. Puffins—Reintroduction—Maine. 2. Atlantic puffin—Maine.
3. Birds—Reintroduction—Maine. 4. Wildlife reintroduction—
Maine. I. Jackson, Derrick Z., 1955– II. Title.
QL696.C42 K75 2015
598.4—dc23   2014040225

A catalogue record for this book is available from the
British Library.

This paper meets the requirements of ANSI/NISO Z39.48–1992
(Permanence of Paper).

10 9 8 7 6 5 4 3 2 1

To the "puffineers" of Maine

Contents

Preface

IT IS EASY TO DESPAIR ABOUT conservation and the environment. If it is not elephant poaching in Africa, it is polar bears losing their Arctic ice. Scores of species of birds, mammals, fish, and reptiles, many with no common names, are marching to the brink of extinction. Despite all the science at hand that human-induced climate change has the planet spiraling into devastating levels of extreme weather and ocean rise, the most powerful government in the world, the United States, still has no comprehensive energy or climate policy. John Muir would have lamented this current state, having once written, "God has cared for these trees, saved them from drought, avalanches and a thousand tempests and floods. But he cannot save them from fools."

Birds are a major witness to the tempests and a victim of the fools. In 2013, the International Union for Conservation of Nature (IUCN), the principal keeper of records on the status of the world's biodiversity, included on their Red List of threatened species 1,300 of the world's 10,000 bird species—a stunning ratio of one in eight. The same list reported that the num-

ber of critically endangered birds had reached an all-time high
of nearly 200.

Leon Bennun, BirdLife International's director of science, policy, and information, notes that bird recovery is slow
because of habitat loss, agricultural changes, invasive species,
and climate change. Seabirds face additional threats, including
rising ocean levels, which can flood nesting colonies; industrial fishing, which depletes forage fish and causes entanglement in fishing gear; marine pollution from acid precipitation;
and oil, plastic, and chemical pollution. Little wonder that seabirds are the most threatened group of birds on earth, with 28
percent of the world's 346 species listed as globally threatened.

Although a few iconic birds such as the bald eagle and
whooping crane have been brought back from the brink of extinction with the help of the Endangered Species Act, the
plight of less well known birds such as the Florida grasshopper
sparrow, Gunnison sage grouse in Colorado, and Kittlitz's mur-
relet in Alaska receive less publicity, and their future is grim.

Although never listed as endangered, Atlantic puffins
were plundered in Maine and Canada in the nineteenth century for food and feathers and were reduced to just one pair on
a Maine island by 1902. This book is the story of the first forty
years of Project Puffin and its quest to bring puffins back to
Maine. The project began with the knowledge that no seabird
had ever been restored to an island where humans had wiped
them out. On a shoestring budget and against daunting odds,
Project Puffin has provided a model for similar conservation
programs worldwide and offers new tools and proof that individuals can make a difference and enterprising conservation
programs at local levels can have larger benefits to species
conservation.

Today, more than a thousand puffin pairs are nesting on

four Maine islands, and this development has led to a flourishing industry called puffin watching. It is ironic that at the same islands where gunners once used eight-foot-long punt guns to blast seabirds fifty at a time, now the sight of even a single puffin is greeted by cries of delight from boats carrying a hundred or more puffin watchers. This change in value is one of the more hopeful signs of our time.

Maine gift shops now profit from the sale of T-shirts, stuffed puffins, and puffin earrings. These symbols help Maine's economy, but they also help remind people that this iconic bird, like a phoenix, has come back to its historic home. The puffins' return is testimony to the possibilities of restoration and the reality that people can not only push species to extinction but also revive lost populations and expand ranges. The willingness to take on active management opens great possibilities for helping wildlife.

On a planet burdened with too many people, waiting for the "balance of nature" to restore lost species will only speed the demise of many additional species. The alternative to stewardship is to passively chronicle extinctions. We believe that the lessons of stewardship learned from Maine puffins have broad applications, not only to saving rare and endangered birds, but also to helping wildlife everywhere at all scales—including backyard habitats.

The return of puffins to the Maine coast is proof that people can, instead of exploiting nature, become responsible stewards for the planet. This inspirational story is not just about bringing a singular bird back to a former nesting island. We tell it in the hope that the tempest calms, the skies clear, and conservation becomes the guiding principle that averts the tragedy of future bird extinctions—and perhaps our own.

In hoping for the restoration of places stripped of native

flora and fauna, the great biologist E. O. Wilson said, "A wilderness of sorts can be reborn in the wasteland." We hope one place of rebirth is on seven acres of jumbled rock six miles off the coast of Pemaquid Point, Maine.

Acknowledgments

OVER THE FORTY-YEAR HISTORY of Project Puffin, many people have played key roles in making this work possible. Some are no longer with us to read these pages, but without them there might be no story at all. They include several mentors: Irving Kassoy, Carl Buchheister, and William Drury, who shared their own enthusiasm for birds, passion for conservation, and belief in Project Puffin. I extend my deepest gratitude to Duryea and Peggy Morton, who offered me my first experiences with Audubon, my invitation to teach at Hog Island, and ongoing enthusiasm from the genesis of Project Puffin to today.

I also thank my very loyal Project Puffin staff, who have together given more than a hundred years of time to this project. Deep thanks go to Rose Borzik, Scott Hall, Ruth Likowski, Susan Meadows, Pete Salmansohn, Susan Schubel, Paula Shannon, and Deborah Wood. It is with my greatest appreciation that I acknowledge the several hundred private supporters and foundations that understand the need for ongoing management and the value of training the next generation of field biologists. I am most indebted to these young people who share my passion for puffins. Since 1973, several hundred interns and island supervisors have followed Kathy Blanchard, the first Project Puffin intern.

Peggy Morton dubbed this hardy group "puffineers" years ago for their dedication to seabird conservation—an endearing name that has stuck even as we have expanded our work to entire seabird communities. Embracing extreme weather, they are compensated mainly by an "aesthetic paycheck" of sunrises, sunsets, and glorious birds to serve as seabird stewards in an Audubon tradition that dates back to the wildlife wardens who halted feather hunters more than a century ago.

Project Puffin's interns have certainly gained inspiration from living among restored colonies of seabirds. But the inspiration goes both ways. The interns continually inspire me by caring so deeply about the birds. They are the greatest hope that imagination and dedication to wildlife is alive and well among today's new generation of conservation biologists. Some of their stories are shared in these pages. The conversations in the book come from my clear recollection. They are not written to represent word-for-word transcripts. Rather, the accounts presented here are retold in a way that evokes the feeling and meaning of what was said and written at the time to provide an accurate essence of the dialogue and events.

This book has greatly benefited from discussions and review of sections of the manuscript from friends and colleagues. Many thanks go to Howard Albin, Kevin Bell, Kathleen Blanchard, Roland Clement, Tomohiro Deguchi, Tom French, Frank Graham Jr., Michael P. Harris, Hiroshi Hasegawa, Brad Keitt, Jeremy Madreios, Mary Majka, Jerry McChesney, Duryea and Peggy Morton, David N. Nettleship, Richard Podolsky, Daniel Roby, Steve Sawyer, and David Wingate.

I also thank my good friend and coauthor, Derrick Z.

Jackoon, fui his research, writing, and inspiration over the past twenty-five years. To provide an even tone to our narrative, Derrick's extensive research and interviews are retold using my voice in telling this story. We both thank our literary agent, Wendy Strothman, for encouraging us to write this book and guiding its path to publication. We also thank Jean Thomson Black, our editor at Yale University Press, for her assistance and for taking a personal interest in puffins. The book has also benefited from the donation of an excellent map by Robert Houston of the USFWS Gulf of Maine Program and photos from the early days of the Project by Duryea Morton. Last, I thank my wife, Elissa, for her help, patience, and encouragement with this book and for understanding my passion for puffins.

STEPHEN W. KRESS
Ithaca, New York

I thank my wife, Michelle Holmes, who asked me to go camping to see the foliage in New Hampshire when we were dating and stayed with me even though my response was, "How long can you look at a red leaf?" I also want to acknowledge my oldest son, Omar, who taught me compassion for the natural world when he was eight years old and demanded that we pick up a wounded blue jay alongside a bike trail in Minnesota and wheel it to an animal shelter. I am in debt to my youngest son, Tano, whose quest to be an Eagle Scout led me to mountains from New Mexico to Switzerland that I otherwise would have never seen. I want to thank my coauthor, Stephen W. Kress, for allowing me into the inner workings, er, burrows, of his life's work and entrusting me with the task of helping him to fledge a story that has changed seabird conservation forever. I would like to acknowledge Les Payne, my editor at *Newsday*, who did not laugh when I suggested my visit to Eastern Egg Rock in 1986, and *Boston Globe* editors Renee Loth and Marjorie Pritchard for seeing puffins occasionally as important as politics. Last and most important of all, I dedicate my share in this book to my mother, Doris, who passed away in 2014 during its completion. She drove me to my first sports-writing assignments for the *Milwaukee Journal*, inspired me by going to junior college the same time that I went to college, and supported my love of journalism despite it being an unusual craft in my working-class neighborhood. When I was a child, some vandals stole flowers out of her humble garden. She ran after them. She tripped and fell on her bosoms. But the vandals never came back. I attribute that to her courage to protect her tiny part of the planet, courage no less fierce than that of any environmentalist.

DERRICK Z. JACKSON
Cambridge, Massachusetts

Introduction
My Passion for Puffins

My passion for restoring seabird nesting colonies began in 1969 while I served as the ornithology instructor at Hog Island Audubon Camp in Bremen, Maine. Hog Island was the signature camp of the National Audubon Society. There, guests learned birding at the feet of such luminaries as Roger Tory Peterson—arguably America's most famous naturalist successor to John James Audubon—and Allan D. Cruickshank, renowned bird photographer and prolific lecturer.

A standard camp field trip is a cruise in Muscongus Bay, circling various islands, including Eastern Egg Rock, among the smallest of the more than 4,600 Maine islands and ledges. At a distance, Eastern Egg, which rises only seventeen feet above the sea, looks like an unremarkable, treeless clump of boulders and wave-washed granite. Its center is a flat meadow with just inches of soil made up mostly of decayed peat and bird guano. The tiny island is dominated by chest-high raspberry and elderberry bushes. But actual berries are rare because the soil is so enriched by bird excrement that plants opt for leaves and stems rather than fruit. Clumps of three-foot-high blue flag are the only notable flowers that poke up from prickly raspberry and nettles. There is no electricity or fresh water. Here, the only running water is the surf that thunders against the rock. From a chance encounter with a book on Maine's birds in the camp library I learned that Eastern Egg was formerly a nesting place for Atlantic puffins, and from that seemingly minor fact grew Project Puffin and my life's commitment to helping seabirds.

The Atlantic puffin's black-and-white plumage, which mimics a friar's robes, prompted eighteenth-century zoologists to name it *Fratercula arctica,* "little brother of the north." Thanks to its brilliant, clownish red-orange-and-yellow bill, the puffin was once known as the sea parrot, though the beak more closely resembles the rainbow bill of a toucan. Many today consider the puffin to be the most endearing bird of the North Atlantic. That was not the case in the nineteenth century.

In Maine, local fishermen and coastal farmers hunted puffins and puffin eggs for food and extirpated them (local extinction) on every US island by the 1880s, save for a pair in 1902 on Matinicus Rock, an island twenty-two miles south of Rockland.

By the time of my arrival decades later, birdlife on Eastern Egg Rock was reduced to a monotony of herring and great black-backed gulls. Magnificent birds in their own right, gulls are among the most resourceful and adaptable animals in the world, eating everything from dump garbage and flying insects to other birds. It may be difficult to imagine, given their twenty-first-century domination of shorelines throughout their vast range, that gulls were slaughtered along with the puffins and nearly wiped out in the late nineteenth century for their prized feathers to adorn fashionable hats in the millinery trade. Herring gulls were so scarce on the Maine coast that they became symbols of conservation and helped to galvanize the early bird protection movement.

Passage of the Lacey Act of 1900 and, later, the Migratory Bird Treaty Act of 1918 made it illegal to kill most North American birds, including puffins and all other seabirds. With bird protection laws in place, enforced at gunpoint by Audubon wardens, adaptable species such as gulls were relatively quick to recover. Indeed, by the mid-twentieth century, gull populations burgeoned due to abundant food plucked from landfills, commercial fishing, and a culture disconnected from the downside of waste.

The great auk became extinct because of these same relentless pressures. Not by intent, but because no one knew the extent of the slaughter and no one cared or had the means to save it. While gulls, eiders, and cormorants made a comeback in response to the end of the hunting days, other colonial birds like the puffin did not. This was in part because, unlike gulls that can rear three chicks per nest and breed almost anywhere there is food, puffins do not breed until they are about five years old, lay just one egg each breeding season, and typically return to breed at the island where they hatched, rather than founding new colonies. Also, young puffins are usually reluctant to nest near gulls, much preferring to nest in existing puffin colonies. Herring and black-backed gulls have the advantage over other seabirds because they feed not just on garbage but also on the eggs and chicks of other birds, including puffins. Therefore, even with wardens watching over the sole outpost of Matinicus Rock, the puffins never reclaimed any of the islands where they had been eliminated.

I could not bring back the great auk, alas, but as I guided

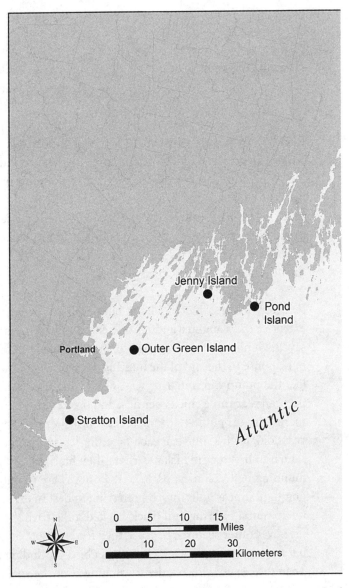

Project Puffin operates summer field stations on seven Maine seabird nesting islands. These seabird sanctuaries provide nesting habitat for thirty-two waterbird species, including most of Maine's puffins and terns.

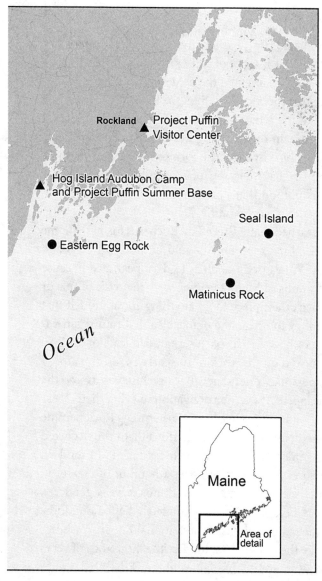

Resident teams of college-age "seabird stewards" protect the birds during spring and summer. Map by Robert Houston, Gulf of Maine Program, USFWS

our Hog Island field trips, I asked myself whether humans had to settle for the local absence of still-existing creatures. In 1916 Theodore Roosevelt lamented the great bird millinery slaughters by saying that their loss was like "the loss of a gallery of the masterpieces of the artists of old time." No artist on canvas, I still began to dream of rendering a new scene on Egg Rock. Could I restore, bird by bird, the former richness of these islands? Could I replace decades of mourning the lost gallery with a new exhibition that illustrated the human capacity to repair damaged bird communities?

Once I learned that Eastern Egg Rock was formerly home to a thriving colony of puffins, I could not look at the island the same way. We caused the birds to disappear. Just because several generations of Mainers either were unaware of this or thought it wasn't their place to do anything about it, I felt that it was time to try to repair the damage so that future generations would not have to accept this degraded state of birdlife. The more I looked on this rock in the early 1970s, the more I declared it a personal inheritance. Because humans caused the puffins to disappear, it was our obligation to bring them back.

Although it is only seven acres, Eastern Egg Rock became in my mind a metaphor of the extinctions that have occurred for centuries under the heel of human expansion. In modern times to be sure, for reasons of size, a handful of creatures, such as whales, elephants, tigers, and rhinos, have engendered global concern before they were completely wiped out. Others, such as the famed pandas at the National Zoo in Washington, DC, evoke the "cuddly" factor. But the thousands of years of the effects of migrating humans and the "F-bombs" on nature (food, feathers, fur, and fat for food and fuel) have caused the elimination over the past four centuries of such magnificent creatures as the New Zealand moa (became extinct in the fifteenth century), Steller's sea cow in the North Pacific (1768), Atlas bear in Africa (late 1800s), and eastern elk in the United States (1896).

The list is so sadly long that, although the dodo bird of Mauritius (1660s) and the passenger pigeon in the United States (1914) became iconic birds in the ledger of extinction, many, many other creatures have passed beyond most human memory. Who but a devoted birder remembers that the last Carolina parakeet died in the Cincinnati Zoo in 1918 or that the last heath hen probably died on Martha's Vineyard in the 1930s?

Puffins were mostly forgotten as a Maine bird in 1969 when I began teaching at Hog Island—even though a small natural colony existed at Matinicus Rock about thirty-five miles away. The more I thought about it, the more I became committed to making a difference and to seeing if I could bring puffins back to Eastern Egg Rock. In a small way, I felt, it would demonstrate that people could restore nature rather than cause extinctions and diminish populations. This idea of human intervention to directly increase biodiversity was a surprisingly novel idea at the time. Famed Harvard biologist E. O. Wilson would later title his masterwork *The Diversity of Life*. I wanted to restore the bird diversity of Eastern Egg. But how?

1

Chasing Skinks

Long before the puffin, there was the skink.

This lizard was my prize of prizes when I was about ten years old, living in Bexley, Ohio, a cozy village embedded within the eastern reaches of Columbus. It was the mid-1950s, a time when families let fourth-graders romp from house to house and disappear for hours in forested parks. Much of my childhood universe occupied what lay under the shaded canopy of Blacklick Woods Metropolitan Park, a twenty-minute drive through farmland from our suburban home. The park was a paradise of some of the least disturbed beech-maple forest and vernal swamps in central Ohio.

My mom, Lina, would drop me off there with a friend, Mac Albin, with nary the concern displayed by today's parents. In fact, I don't know whether to laugh or shake my head in sadness when I read today of the struggle of families to let children explore these woods. In a 2002 article, the *Columbus Dispatch* wrote how parents and grandparents watch their children play at Blacklick, concerned "that the girls are never completely safe. Not in today's world, where a child can be abducted from a front yard or bedroom." The article quoted Kathy Double of Reynoldsburg, a thirty-six-year-old mother who runs a childcare center for her two children and four others. She said, "When she takes the playgroup to Blacklick the rules are 'If you can't see me, I can't see you. I pick a spot in the park where I can see the whole play area. I follow that old mafia rule: Keep your back to the wall and watch for your enemies.'"[1]

There were no walls for Mac and me as these woods became my best friend. It started in fourth grade when Steve Albin, Mac's older brother, invited Mac and me to come with him to Blacklick for the Saturday morning "Junior Explorers" program. At Blacklick, I found trees, ponds, geology, and birds so interesting that Mac and I asked our parents not to pick us up until hours after Explorers ended so we could ramble on our own. We pleaded with them to take us to the park and drop us off even when Explorers was not in session. The moist swamp forest and vernal pools made Blacklick particularly rich with reptiles and amphibians like wood frogs, spring peepers, and salamanders. Mac and I were never disappointed. Even the tiny fairy shrimp that emerged in the vernal pools after the first spring thaw were cause for getting soaked. The park naturalists were so taken with how happy we were—wet and muddy—that they invited us to help them maintain a trailside collection of native animals.

Our bonus for helping them clean the cages was the gift of snakes, turtles, and even an occasional raccoon for our own backyards. My dad, Herman, who owned a business reconditioning burlap bags for potatoes and plant nurseries, was handy making furniture, and he helped me build terrariums and showed me how to build cages. All that Mom the homemaker asked was that I keep my "zoo" in the backyard or basement. When salamanders and garter snakes got loose in the house, she acted alarmed but ended up using the incident as jolly material at her next canasta game. I began to assume that she vicariously enjoyed my menagerie.

I became especially enchanted with the five-lined skink. They were like no other lizards we encountered, elusive and confined to just a few locations in the park. They were like tiny *T. rexes*, terrorizing beetles, grasshoppers, crickets, and spiders. The young skinks, just three to four inches long, were especially appealing for their neon-blue tails. When we found one sunning on a dead stump, we would admire it for a moment, then pounce on it for capture. But skinks have evolved an amazing method of self-preservation in a world where their predators include a long list of sharp-eyed creatures, including herons, owls, hawks, jays, snakes, skunks, raccoons, and foxes. If it grabs the skink by the tail, the predator soon finds itself holding nothing more than that wiggly tail.

Meanwhile, the skink sprints off to the nearest patch of poison ivy or other ground cover to grow another tail and live another day. The greatest disappointment was to lose the prize and be left holding this wriggling, Day-Glo tail. Outdoor ethics were unknown to me at the time, and, oddly, I don't recall the otherwise sage naturalists talking to us about wildlife conservation.

Another favorite activity for Mac and me was seining up little fish from local creeks for aquariums. Curious to see if fish could distinguish colors, I rigged a fifteen-gallon aquarium with electromagnetic feeding cups on each side of the tank. My idea was to draw the fish to the four corners of the tank in response to a bank of differently colored lights. Each color was paired with a different side of the tank. At first I paired food with light and trained the fish to visit different stations in response to the colors. After the conditioning was complete, the lights alone would send the fish to the appropriate corner of the tank. I had great plans to use the system for developing an intelligence test for fish and to begin a ranking of different species based on the time it would take them to condition to the lights. I thought the experiment was working until someone asked me, "How do you know it isn't the brightness rather than the color they're seeing?" That taught me at an early age that colorful experiments often wind up shaded in gray.

Other people asked me questions that piqued my interest in nature. On a spring day in 1954, at Montrose Elementary School in Bexley, my fourth-grade teacher, Mrs. Reed, spotted a bird down below our second-floor classroom window. She drew the class to the window and asked us, "Who can identify that brown bird poking at the ground?"

I ran to get the dog-eared Golden Guide to birds she kept on her shelf. When the robin-sized bird rose into the air, displaying a white rump and a flash of yellow on its underwing, it was all I needed for a correct answer. That was enough for me to thumb to the drawing of a northern (then called yellow-shafted) flicker. Hence my first success in bird identification! I begged for Golden Guides as birthday and Hanukkah presents, and many nights, my mom would come into my bedroom to make sure I was asleep only to find the light on and a Golden Guide at my pillow or dropped down on the floor under my dangling arm.

Dad acknowledged my enthrallment by bringing home box turtles he found crossing the road while visiting farmers on his bag sales trips. Dad worked long hours, brought home a briefcase of work, and returned to the office on most Saturday mornings. When he did take Saturday off, our best opportunity for one-on-one time was fishing. He was not especially fond of fishing per se—few Jewish businessmen like him bothered with it. But perhaps because his hard-working father, a deli owner, died of pneumonia when Dad was just ten, he was happy to spend time with me, taking me fishing along Big Walnut Creek and other small rivers where we could soak a line.

Because he was not cut out of the L.L.Bean mold, Dad was often impatient even as he tried to please me. After a few casts, there would be the usual proclamation, "There's no fish here. Let's go." Although I was happy to pull in a sparkling sunfish or prickly backed catfish, Dad's time meant more than the actual fishing. One day, his hand was pulled into a printing press at the bag shop. He lost a finger, and a thumb was crushed. When I visited him at the hospital, I cried, but it was not because he was hurt. It was because our fishing date was canceled.

And further truth be told, I was always more interested in amphibians than fish. Many of our sessions started with trying to hook something on the line, but when fishing was slow, I happily turned to pouncing on frogs. The first day I experienced a sense of loss for an animal was on a fishing trip. I was thrilled to find a box turtle sitting by one of our favorite fishing holes. But when I picked it up, its head was gone. Somebody had recently cut it off with a clean slice, leaving just a stump where its head had been. Was it sacrificed for bait? Or was this just a cruel joke? I was devastated.

That day I began to think about my own impact on nature. I thought of all the salamanders I had kidnapped from moist trails and how they had dried up or overheated in my jars. Even as a child, I began to see that I too was a predator and I was having a negative effect on animals.

As these darker thoughts were dawning on me and I was outgrowing Junior Explorers, Mac and I heard there were some serious birdwatchers willing to take a couple of kids to look for hawks and owls. We immediately signed up. It was another late 1950s dream that would be unthinkable today—a stranger with wheels offering to take us to the woods!

Irving Kassoy was no casual community volunteer. He was a member of Columbus's elite Wheaton Club. Founded in 1921, the club was loaded with Ohio State University professors, state natural history museum officials, and writers. It was named after naturalist John Maynard Wheaton, a Columbus physician who was born in 1840 and died of tuberculosis at forty-six. Legend has it that Wheaton's love of nature came about as a child because he kept himself outdoors perpetually seeking fresh air for his diseased lungs.

Kassoy, born in Russia in 1904, was brought by his family to the United States as a toddler and grew up to become a jewelry salesman in Manhattan. Before moving to Columbus, he had established his own business in 1936, selling supplies to diamond dealers. The firm bearing his name still exists in New York City and Long Island; Kassoy, Inc., specializes in microscopes and magnifying loupes for jewelers that compete with the lenses of Zeiss and Bausch & Lomb.

But it was not glistening diamonds dangling from the necks of the rich and famous that stirred Kassoy's deepest passions. It was the darkest corners of woods and buildings. He became such an expert on barn owls that a writer for the *New Yorker* followed him around for a story in 1936 titled: "Owl Man." The article described him as "a slim little fellow of 32, very pale and somewhat bald. His chin is pointed and his skin is drawn taut from his cheeks up to his wide forehead. You couldn't help noticing his grave eyes because of the thick glasses he wears. Even to one not knowing the nature of his hobby, Mr. Kassoy's general appearance would be hauntingly suggestive of owls." Kassoy was so good at his craft that the publicity manager for the Audubon Society declared, "The National Association of Audubon Societies gets all its owl reports from Mr. Kassoy."[2]

The barn owl was Kassoy's favorite because that species was a year-round city resident. In 1932, he heard about a barn owl nest in the ventilator of an old mansion in Pelham Bay Park in the Bronx. The owl family produced five chicks out of six eggs in the spring of 1933. The next year Kassoy convinced the city parks department to give him a set of keys to the mansion so that he could set up a one-man midnight-owl vigil. He placed a box at an opening of the ventilator, lit the space with a battery-powered light, and fixed into place a one-way glass plate that would let him see into the nest.

Kassoy once took Roger Tory Peterson to the mansion. As Peterson later recounted in *Birds over America:* "The caretaker believed the house was haunted and I am sure he thought Kassoy was quite mad, sitting up there night after night by himself. . . . Huddled there were five of the most grotesque owlets I have ever seen, like little monkeys in fuzzy bedclothes, with white caps pulled over their ears. . . . When I called the owlets grotesque, he was indignant. To him they were beautiful babies. . . . Night after night, 200 or more, he hunched over his box like some immobile Buddha, his face, reflecting the wan light, the only thing visible in the blackness."[3]

Kassoy and Peterson belonged to the Bronx County Bird Club, arguably the most elite collection of amateur naturalists in the nation. The club started in 1918 with eleven-year-olds thumbing through a bird guide used by a Boy Scout for his birding merit badge. By 1924, when most of them were no older than seventeen, the club had blossomed into a formal organization. One of its purposes was to participate in the Audubon Christmas Bird Count. Kassoy and Allan Cruickshank were two of the nine original members. Another was Joe Hickey, who would take his interest in birds to the University of Wisconsin, where he studied under the father of wildlife management, Aldo Leopold. Hickey wrote an acclaimed bird-watching guide, and his pesticide research in the 1960s was crucial in the banning of DDT. Part of Hickey's concern about DDT came from club cofounder Richard Herbert, who charted the disappearance of peregrine falcons from cliffs along the Hudson River. Peterson joined in 1927. It was remarkable that so much brilliance could come from a bunch of boys, who, as Peterson later remarked, "were addicted to the Hunts Point Dump." Where the rest of New York City's populace saw foul refuse, they found four snowy owls feasting on rats.[4]

By 1950, Kassoy had sold off his jewelry supply business and moved to Columbus so that his wife could be near her family. By weekday, he was an upholsterer. On Saturdays, armed with binoculars and spotting scope, he became my birding superhero, arriving at my front door at dawn in a beige Plymouth station wagon to pick up Mac and me as my parents slept.

We usually first headed off to a local diner for bacon, eggs, and toast. Often there would be another member or two of the Wheaton Club in the car, such as Milt Trautman, the undisputed authority on the fishes of Ohio, or cigar-smoking local artist David Henderson, a great admirer of Francis Lee Jaques, painter of wildlife habitat dioramas at the American Museum of Natural History in New York. Henderson encouraged my own interest in sketching and watercolors of birds and frogs and later gave me several of Jaques's books that were inscribed to him.

For four years, when I was thirteen until sixteen, these men would take me, fall through spring, on their search for rough-legged hawks, short-eared owls, and barn owls, collecting notes on horned larks, white-crowned sparrows, and golden plovers along the Ross-Pickaway County Line Road in southern Ohio.

While other kids played sports, I tromped through corn stubble and grassy fields counting short-eared owls. During the day, these owls roost on the ground, so we'd line up side by side, then spread ourselves out in a long, straight line and walk across the field to flush the birds from their communal roost. We would usually see twenty or more of the graceful birds. In late afternoon, we would sometimes see the owls hunting over the fields. Occasionally, we saw them clap their wings under their bodies as part of a courtship display.

In time, I became something of a nephew to Kassoy. He had a cute, curly-haired daughter named Laura on whom I had a secret crush. But she wanted no part of her dad's bird-watching obsession and blamed birds for her parents' divorce. The fact that I was a birder pretty much killed any chance I had with her. Perhaps I was a substitute for the daughter Kassoy would have preferred riding shotgun.

During one of our many car rides, Kassoy shared what he knew about the roots of American birding. Ludlow Griscom, who died in 1959, right about the time these Saturday trips started, had been a friend and mentor to the boys of the Bronx County Bird Club. Griscom, born in 1890, is often credited as inspiring modern bird-watching, since he was among the first to demonstrate that it was possible to identify birds with field glasses instead of blasting them with a shotgun. Legend has it that an old-school shotgun bird collector dared Griscom to prove that he could identify a warbler—possibly in Central Park— on field markings alone. Griscom saw a warbler in a tree and declared it a Cape May. The old-school birder gunned the bird

down to the ground, where they both saw that it was indeed a
Cape May.[5]

Griscom attracted the attention and mentoring of bird
preservationist Frank Chapman, the visionary behind the first
Christmas Bird Count in 1900. In turn, Griscom was taken by
the genius of others younger than he. By the early 1920s, he
became an assistant curator at the American Museum of Natural History in New York. In his spare time, he became a devoted mentor to Irv Kassoy and the Bronx County Bird Club.
Kassoy recounted that Griscom would often cross-examine
the boys "ruthlessly" when they claimed to spot unusual birds.
Under Griscom's tutelage, the "Bronx boys" were the first group
of birders, as Peterson wrote, to "break through the 'sound
barrier' of 100 species on a Christmas list." This was a game,
admittedly, but a useful game, a test of skills and part of the
basic training of any competent ornithologist. Griscom was
the high priest of this new cult of split-second field identification. Peterson has said his own field guides, which would
become the most famous in the nation, were "profoundly influenced" by Griscom.[6]

And there I was, sitting in the car with Kassoy, and my
Peterson's Field Guides! Peterson was pretty close to God to
me. Then one day, RTP himself came to Columbus as a speaker
for the Audubon Wildlife Film Series, and Irv introduced him
to me as his friend.

I was most assuredly at the feet of God. I became a regular attendee of the Audubon series in Columbus during my
high school years in the early 1960s. There, at the back of the
natural history museum on the Ohio State campus, I noticed
brochures for Audubon's summer camps in Maine, Wisconsin,
and Connecticut. Even though the camps were for adults eigh-

teen and over, and I was sixteen going on seventeen, I wrote to each of the camps, hoping they had a summer job for a nature boy. I thought I might have a chance since my omnipresence at Blacklick Woods had landed me a job the previous summer tending the park's trailside zoo.

Duryea "Dur" Morton gave me that chance in the summer of 1963 at the Audubon camp in Greenwich, Connecticut, where teachers would take weeklong nature study programs. I did not know it at the time, but he hired me for exactly the same job for which he was hired by Audubon out of the army at the end of World War II: student assistant dishwasher. He eventually rose to be National Audubon's vice president for education, overseeing Audubon centers and camps. He liked to joke, "All good men start as dishwashers."

Dur would usually gather the student assistants together before the arrival of a new group of campers to offer a pep talk. At one such gathering, he explained that humans were the only species capable of destroying other species, and therefore we also had the responsibility to be caretakers. It was the first time I had ever heard the concept of stewardship, amplifying my unease about that headless turtle and the unfortunate sun-baked salamanders of my youth. For the first time, I started thinking less about how to make creatures part of *my* environment and more about observing and even participating in *their* world.

That summer, my student assistant project focused on bullfrogs and their vocalizations. One night, I noticed that every time an airplane flew over, the bullfrogs would start singing. Night after night, I went to the edge of the lake after finishing my enormous piles of dirty dishes. Dur and my project adviser, Herb Krause, a renowned chronicler of the harsh life of the prairies, were so impressed with my experiment that word of it made it to Audubon president Carl Buchheister, who submitted it to *Audubon Magazine*. To my amazement, the magazine ran the article. The acceptance letter hung on my Bexley bedroom wall for many years. For lay consumption, my title of "Factors Affecting the Vocalizations of Green Frogs *Rana clamitans* and Bullfrogs *Rana catesbeianna*" became "Bullfrogs Sing along with Jets."

The following summer, I returned to Greenwich, this time as "lodge keeper"—basically maintenance of a motel-like dorm for the Audubon Camp. Buchheister visited the Greenwich camp frequently during those two summers. One of his first stops was always in the kitchen to acknowledge the important work of the student assistants. Often during his visits, he would gather the staff at the Mortons' home to share stories of his experiences with bird conservation. He was an impassioned storyteller, and I was especially entranced by his accounts of living on remote islands off Maine and Mexico. I didn't realize it at the time, but his enthusiastic stories had already instilled in me a sense of wonder for seabirds, remote island lighthouses, and the Maine coast. After this second year, I asked Dur if there was an opening at the Audubon Camp on Hog Island. Unfortunately, Dur explained, there was a long-entrenched teaching team on Hog Island and no openings.

Instead, Dur offered me an opportunity the next two summers to join the staff of a sister camp, Hunt Hill Audubon Camp, located in northwestern Wisconsin near Rice Lake. By day, I was officially the assistant birdlife instructor. At dusk I continued to delight in frogs, spending many evenings at Hunt Hill's twin lakes, where I studied the ability of green frogs to return to their home shore.

After dark, I would catch a large frog along the lakeshore and tie one end of a three-foot-long thread around its waist and the other end to a balloon. I would then paddle my canoe to the center of the lake. Once there, I'd place the frog and balloon in a screen-and-fabric funnel that was weighted at the pointed end with metal washers so that it would slowly sink in the water. Then I would back off to watch at a distance. Intermittently I would shine my flashlight on the balloon to gauge where the frog was in relation to its home shore. Most frogs, it turned out, had an uncanny ability to begin moving in the proper direction right away.

When Dur came to visit Wisconsin in his role as Audubon's national camp director, I asked him to join me one evening for a different experiment. We walked to the end of the barn that had been converted to a dining room. There I had hung huge plastic sheets and set black floodlights behind them to create a funnel trap for moths. I was trying to see if weather patterns were associated with how many moths came to the sheets and which species would be most likely to show up on a given night. It must have been a good night when I brought Dur down, because all the moths in the neighborhood appeared to have congregated on the sheets, with hundreds of silhouetted wings of all shapes and sizes.

After two summers in Wisconsin, Dur had a novel idea. An Audubon supporter, Dorothy Eidlitz, had created a combined art and nature center called Sunbury Shores Arts and Nature Centre at her summer home on the shore of the Bay of Fundy in Saint Andrews, New Brunswick. Each year she hired an artist and a naturalist to teach at the center. Dur's description of meadows at the edge of the sea, spruce and fir forests, and an intertidal zone that dropped up to forty feet sold me on the naturalist job.

Exploring tidal pools with children and teaching bird

biology to adults was pleasant enough, but I ended up living for the weekends, when I explored more remote habitats with Mary Majka. Mary came to Saint Andrews to take my birding class and stayed on to volunteer. I was twenty-one; she was in her early forties. In short order, it became clear that she was my "student" in name only.

Majka was born in Poland as Emilia Maria Adler to a wealthy family, but in her early twenties, when the Nazis closed in on Poland, she fled to Czechoslovakia. She was arrested and sent into forced labor camps for two years. After the war, Mary attended medical school in Austria, where she met her future husband, Mietek Majka. They moved to Canada in 1951, where Mietek (Mike) became a physician and Mary began raising a family, abandoning her own medical career. Interested in nature since childhood, she became passionate about conservation after reading Rachel Carson's *Silent Spring* in 1962.

Mary Majka is especially noted for starting a campaign to protect one of North America's most important staging areas for shorebirds in New Brunswick, near her home, which coincidentally is known as Mary's Point. She began buying land in the area, building trails, and inviting community support with education and volunteerism. Her efforts resulted in the creation of the Mary's Point Western Hemispheric Shorebird Reserve. She became a Canadian Rachel Carson, earning her own nature television show and in 2006 receiving the Order of Canada, the nation's highest civilian honor. She refused to believe that anything was impossible—a belief that I have long shared.

It was through Mary that I first saw puffins. One weekend I drove with her and her family down to Cutler, Maine, just over the US border, where we met Purcell Corbett, a Maine coast seaman who used only his watch and compass for navigation to Machias Seal Island. In classic down east humor, he loved to joke when people asked when the fog would lift. He insisted that we had best get used to it because fog is made down east. He may not have known why—but he was right— fog is indeed produced offshore where cold currents from the Arctic mix with the warm water of the Gulf Stream. When wind blew from the east, "thick" fog draped the coast for days at a time.

An hour's boat ride later, with Corbett using yet another old-school navigational tool—his ear—he picked up the calls and then the flight of terns heading toward the island with fish for their chicks. Later, after the terns showed us the way, we heard the booming voice of the Machias Seal horn. Like the opening scene of a staged drama, the fog lifted to unveil a vast intertidal slope of seaweed and a treeless jumble of rocks, crested by a manned lighthouse. On the map, Machias Seal Island is in US waters. But in the early 1800s, the Canadians took far more interest in it because it was a major wreck zone for their ships. They built a lighthouse and homes for lightkeeper families on the island in 1832. Canada officially expropriated Machias Seal in 1912, declaring it part of New Brunswick.

The lightkeeper when we visited was Jack Russell, who made money on the side offering overnight lodging in his home for birders. Jack's wife, Rita, warmed us up with a grand seafood chowder. Then Jack took us out onto the boulders for the birds.

It was my first visit to a seabird colony, and I found it dazzling from the first moment ashore when we were attacked by a swarm of Arctic terns that nested near the shoreline and along the walkway to the buildings. We waved stems of angelica overhead to ward off the attacking birds that were intent on defending their fluffy chicks hidden in nearby patches of vegetation.

By the late 1960s puffin numbers at Machias Seal Island had recovered to about two thousand birds. My first puffin views left an unforgettable memory. Photo by Derrick Z. Jackson

Razorbills and puffins were more demure, perching atop the bird blinds, buildings, and high rocks. My eyes could not have been wider when Jack pulled out adult puffins and chicks from their rocky burrows and storm-petrels from their peaty homes.

At night, we searched under the light tower for courageous puffin fledglings that were working their way toward the sea. Sometimes the chicks, confused by the lights, marched instead to Jack's porch and gathered under the doorsteps. Jack loved to tell the story of one very dark night when the fog was so thick he could slice it with a knife, and he heard a tapping at the lighthouse door. He opened the door cautiously to see who had knocked and found nothing but the fog, until he happened to glance down at an eight-inch-tall puffin that had been pecking at the door.

There were perhaps two thousand puffins frequenting Machias Seal Island in those years, an amazing number, considering how heavily hunted they had been in the nineteenth century. Shooting had reduced their numbers to around sixty birds in the 1880s for two reasons. First, the lightkeepers, hungry for fresh meat after a long winter, considered them fair game. Second, seabirds sitting on buildings were a threat to the precious island water supply that was collected off the rooftops. Keen to protect their water supplies, hunters set their rifle sights on birds that were anywhere near the roofs.

I returned for a second summer at Sunbury Shores in 1968 and continued my weekend seabird adventures, making more trips to Machias Seal, to Bonaventure Island to view gannets, and to Kent Island near Grand Manan to see the large Leach's storm-petrel colony. This taste of seabird life made me ask Dur again if I could come aboard the next summer at the flagship Hog Island Audubon Camp in midcoast Maine.

A few years earlier, in 1964, I had become a zoology major at Ohio State University in Columbus. Ohio State was the logical school for me. Dad and my brother were both OSU grads, and there was no thought of shopping for schools outside Ohio. Plus, Ohio State's zoology program had a natural history track. I enrolled hoping to be a naturalist someday like my heroes at the Columbus Metropolitan Parks.

Still, OSU did its best to make biology dull. Most of my early classes were taught on closed-circuit TV channels with the professor cued up in a remote studio. I mucked through most of them but aced anything ending in "ology." My savior professor was E. E. Good, who introduced me to the Field Biologist's Forum, of which I ended up becoming president. In the basement of the botany and zoology building, where the forum met, were dozens of aquaria, all bubbling and active with fish, frogs, and salamanders on temporary leave from the Ohio countryside. Despite the vast population of Ohio State, forty thousand students at the time, I was very at home in that room.

After graduating from Ohio State in 1967 with my BSc degree, I began work on a master's with Good. After completing the necessary classes, my academic life came to an abrupt end with the escalation of the Vietnam War. Graduate students were no longer deferred from the military draft, but teachers still were. Not eager for combat, and with my most violent act thus far in life being separating a skink's tail from its body, I was fortunate to land a job in Yellow Springs, Ohio, as assistant director at Antioch College's Glen Helen Outdoor Education Center. In this capacity, I took up residence and found sanctuary from the draft at Glen Helen from 1968 to 1972.

In my new position, I started an ornithology course on the Antioch campus and trained elementary school teacher–naturalists. While working at the Glen, I obtained an injured red-tailed hawk that I dubbed Big Red from the nearby Aullwood Audubon Center. This bird and a bum-winged turkey vulture named Buzz were the first residents of the Glen Helen Raptor Center, which is today a widely recognized center for the rehabilitation of and education about Ohio raptors.

Despite being uprooted from my graduate program at OSU, I began to collect data for my master's thesis, titled "Effects of Human Disturbance on Nesting Success of Upland Thicket Birds." The area around Yellow Springs was dominated by shrubland that grew over abandoned farm fields. It became home to catbirds, thrashers, blue-winged warblers, chats, cardinals, and towhees. The children who took part in the center's programs were the perfect source of "human disturbance."

I measured the hours of kid disturbance as the children were led through the thicket as part of Glen Helen field trips. I searched for nests and plotted territories on maps of singing male birds. My hunch was that I would find fewer nests in the disturbed habitat than in undisturbed controlled areas. To my surprise, I found more birds nesting in the disturbed habitat. Here some birds even built their nests on trees hanging over the trails or at the edge of trails.

I realized that the children were to cardinals what alligators are for nesting birds in southern swamps. Wading birds often nest in trees overhanging alligator ponds because the gators keep away the raccoons and other predators. This may explain why many birds, like robins, nest on porches and breezeways and under awnings, or on tree branches close to a house, because predators are often wary about being near human activity. The study taught me to question the expected and prepare for surprises.

2

Ghosts of the Gallery

I t's impossible not to be stunned by your first view of Hog Island. It sits just offshore, the first of dozens of islands that pop into view as you arrive at the end of Keene Neck Road on Maine's midcoast, about forty minutes' drive north of bustling Booth Bay. The heavenly scented spruce-fir forested island was a gift to Audubon by Millicent Todd Bingham, who inherited it from her mother, Mabel Loomis Todd. Todd and Bingham are known for their role in editing Emily Dickinson's poetry, some of which they brought to a rustic camp built in the flavor of Henry David Thoreau's Walden retreat. To advance the idea of encouraging teachers to include nature study in their classroom and schoolyard lessons, Audubon founded the Hog Island Camp in 1936.

During the decades before my arrival, there was a long tradition of Audubon Camp field trips to bird islands in Muscongus Bay. The excursions went to Wreck Island to observe the great blue heron colony and to Western Egg and Eastern Egg Rocks to see gulls, storm-petrels, and guillemots in their nesting habitat. But in 1969, my first year teaching at Hog Island, a new and cautious camp director, Herbert "Doc" Houston, deemed that landings on the bird islands were so risky that he banned shore excursions to the Egg Rocks for the next two summers. There was no particular reason given, no examples of serious injuries suffered by campers. It was frustrating that all we could do was circle the islands for long-distance views of the wildlife.

I was particularly frustrated because it was during Houston's tenure that I spotted a book in the Hog Island Camp library. It was a 1949 work titled *Maine Birds* by Ralph S. Palmer. The book was a summation of all the relevant literature at that time about the state's birds and a repackaging of raw data gathered between 1894 and 1943 by Arthur Norton, curator of the Portland Society of Natural History. No historical review of the birdlife in Maine had been written for forty years, and Palmer had filled this vacuum.

Piqued by my puffin trips to Machias Seal Island with Mary Majka, I excitedly flipped the pages of his treatise for any historical evidence about puffins. I was not surprised to find an account of Machias Seal. But I was shocked to learn that until the 1880s puffins flourished on many other islands, some not far from Hog Island. Palmer's most descriptive account was from Seal Island, located just six miles east of Matinicus Rock. In a retelling of a 1923 report by Arthur Norton, Palmer wrote, "This is the main breeding place of the Puffin in Knox County. The birds were killed for food, and during the 1850s, parties visited in the evening and spread old herring nets over the rocks to capture the birds as they came forth in the morning. By 1886, the colony had been reduced to about 25 or 30 pairs. 'Their final extermination was probably effected the following year by the milliners' agents who carried on a most destructive season's work at that place.'"[1]

But I was completely blindsided by this passage: "Western and Eastern Egg Rocks in Muscongus Bay Norton . . . reported that the Puffin bred on these rocks prior to 1860 and that there were 'considerable numbers' still on both rocks in the late 1870s. He pointed out that they were much reduced by shooting in the early 1880s, leaving only five or six pairs on Western Egg Rock by 1885. He saw birds and an egg that had been taken there that year. During the next two years, the last of the birds disappeared from the place."[2]

Those six words, "the Puffin bred on these rocks," changed my life. These were the same islands that Audubon campers had circled for decades and landed on in most years, just eight miles from Hog Island. But as far as I knew, every one of those trips was conducted without this knowledge. It was a secret waiting for my discovery. From that moment, I ceased looking at Muscongus Bay for what it was in 1970. Once I realized how diverse the bird community on the Egg Rocks had formerly been and knew that puffins had nested on these islands, they suddenly transformed in my mind from vibrant gull colonies into diminished, monotonous reflections of their former selves. The granite boulders seemed somber without the flight of puffins and terns.

The sadness of this new vision and the deep sense of loss triggered intense curiosity within me. Almost like a tidal surge, all my youthful and collegiate tinkering with fish feeding and lights, frog calls and jets, frog navigation and balloons, moths and black lights, and screaming children dashing through upland bird habitat flooded back to me. It all made me ask, "What if?" If only there was a way to bring the puffins back!

My wonder was fueled by the knowledge that restorations of certain raptors were already under way. Two years before, Joe Hickey, the founding member of the Bronx County Bird Club who had become a wildlife management researcher in

Wisconsin, copublished findings in *Science* magazine that
agreed with emerging research on DDT and concluded that
this chlorinated hydrocarbon pesticide developed to kill agricultural pests and disease-bearing mosquitoes had found its
way into the diets of raptors and other birds.[3] Eagles, osprey,
and peregrine falcons were the most obvious victims. Their
diet concentrated the poison, which caused them to lay eggs
with such thin shells that they were crushed under the incubating parents. This research led to a ban on DDT in the United
States in 1972 and set the stage for the rebounding of these
majestic, fearsome birds that rule atop the avian food chain.

As I started thinking about how to restore puffins, I suddenly
realized how the Egg Rocks were strikingly similar to Machias
Seal Island, where I saw my very first puffins with Mary Majka.
Both shared an interior meadow and a boulder-covered perimeter. My head began to fill with images of puffins prancing
atop those boulders or buzzing ashore with their rainbow
beaks loaded with gleaming fish for their chicks. During the
summer of 1970, I began considering how I might influence
the colonization process and encourage puffins to reestablish
a colony on one of the Egg Rocks. I wasn't sure which of these
islands or exactly how I would achieve this, but I was burning
with the vision of reviving a puffin colony at one of these former nesting places near Hog Island.

By the end of the summer of 1970, I had my first draft of
a plan to "Re-Establish Puffins in Muscongus Bay." Dur Morton thought enough of it to send it along to Roland Clement,
then Audubon's vice president for science, for his consideration for possible funding.

My obsession continued that fall and throughout the
1970–1971 academic year after I returned to Antioch College

in Yellow Springs, Ohio, to continue teaching ornithology and training teacher-naturalists in Glen Helen, but these were tumultuous times at Antioch, a liberal community deeply troubled by world events. The previous spring of 1970, President Richard M. Nixon had announced that the United States would invade Cambodia and students were shot at nearby Kent State University. Protests for civil rights raged, and sympathetic students closed down the Antioch campus and even the trails of Glen Helen. Through it all, I was obsessed with puffins.

I read everything I could find on the topic, starting with Ronald Lockley's 1953 *Puffins.* At the time, it was about the only reference on puffins, gleaned from his years of research on the Welsh islands. I learned that puffin parents raised their chick underground, bringing it whole fingerling-sized fish that they dropped at the chick's feet. According to Lockley, at a certain point the parents stopped feeding the chick, which led to its hunger and eventual fledging to look for its next meal.[4] This abandonment idea, it turns out, is not true in most years and may happen only when food is in short supply.

Now it is known that puffin chicks typically fledge when they are about six weeks old. Using cameras with infrared light tucked deep into a puffin's burrow, we have discovered that puffin parents do much more than deliver food. They brood the chick and preen its feathers almost until it fledges, usually at about forty-two days old. When food is scarce, puffin chicks stay longer in the burrow. In general, puffin chicks fledge when they are fully feathered with few down feathers clinging to their waterproof plumage. Although Lockley did not have all the facts straight, the key parts of the life history really got me thinking. I reasoned that, if I could deliver whole fish as the parents do, then the chicks would head to sea on their own, which was similar to what happens in nature. At the time no one knew how long puffins stay at sea, and it is still not known where they go or what they eat. But I concluded that I didn't need to know all the answers before beginning.

Lockley's sketchy description of puffin chicks was all I needed to hatch a plan. This information made me wonder if I could take puffin eggs or chicks from one island and move them to another before they learned the meaning of "home." This was similar to the plan that was in its infancy at the Cornell Lab of Ornithology, where Professor Tom Cade led a team that had just begun raising peregrine falcon chicks with the intent of restoring populations where they had been lost due to DDT.

When I went back to Hog Island in the summer of 1971, I was greeted with the news that Houston, whose restrictions made for two uneventful and forgettable summers, had moved on. I happily forgot them even more when I learned that his replacement as camp director was my mentor, Duryea Morton, who immediately resumed the bird island landings. We shared the same Audubon camp beginnings as dishwashers, and I began to admire Dur even more than ever for how steeped he was in the camp's history. I quickly learned that a surprising amount of that history involved puffins.

The Hog Island Fish House was the setting for evening programs, storytelling, and occasionally spoofy camp antics. Dur recalled one such special evening in the Fish House after Carl Buchheister and Allan Cruickshank and their respective wives, Harriet and Helen, had just returned from a week on Matinicus Rock. Located twenty-three miles off the Maine coast south of Rockland, the island had a lighthouse made famous by the heroism of Abbie Burgess, a lightkeeper's daughter who kept the lights on through weeks of fierce storms and dwindling food for the family when her father could not get back from his supply runs to the mainland.

Just as Burgess once wrote, "It has always seemed to me that the light was part of myself," so Matinicus Rock's birds became part of the Buchheisters. The island was the only one west of Machias Seal where puffins still hung on. The population was down to one pair in 1902, when Audubon-paid wardens began keeping away hunters. Over time the puffin population slowly rebounded enough for the Buchheisters and Cruickshanks to make a 16 mm film of the birds to show to campers on Hog Island. At the end of the film, Harriet and Helen burst into the hall in puffin headdresses. They pranced down the aisle imitating puffin behavior. They bowed to each other, billed, and waddled about in classic camp spirit.

With encouragement from my Audubon mentors, I began writing to other seabird researchers in the United States and abroad to gather opinions about puffin colony restoration. One of my first letters was to Ralph S. Palmer, author of the book that first inspired my thinking. Palmer, born in 1914, was still at the New York State Museum when I began to contemplate puffin restoration. I assumed that, with his keen interest in Maine birds and because he had a summer home in nearby Friendship, he would take an interest in my growing enthusiasm for seabirds. I also assumed he would be interested because his curiosity seemed insatiable. In 1962, he published the first volume of *Handbook of North American Birds,* which focused on diving birds, and he possessed a personal mammal collection of more than 1,700 specimens and a vast library of rare books on wildlife around the world, as well as the natural world of Maine, Native American history, and North American explorations in the 1700s.

When he was a college student, Palmer traveled to Russia. Even though he did not read the language, he came back with suitcases of books on Russian ornithology. According to legend, he made space for the books by leaving most of his clothing behind. He was known in New York for his encyclopedic knowledge of zoology.

I assumed I would receive a sympathetic ear from an ornithologist who quoted old reports of puffins in his tome *Maine Birds* such as the Seal Island entry, where "their final extermination" was due to "a most destructive season's work" by gunners. I assumed that someone who cared about the puffin's destruction would care about an attempt at restoration.

I was pleased to receive his response in a letter mailed from nearby Tenant's Harbor. But I was shocked to read that he thought that my idea was a stunt and a waste of time and that anyone who wanted to see puffins should go to Iceland.

At first I thought he must be joking. But he was serious. If this was the response from the man who had the most detailed knowledge of Maine puffins, I had to consider whether my idea was hopelessly flawed. Fortunately, I had the encouragement of my mentors within Audubon. Still, I wondered how many others would share Palmer's view. To counter critics like Palmer, I realized that I needed to know as much about puffins as possible, especially the history of their demise and current threats such as predators and ample food supplies.

The documentation of puffin slaughter in the Victorian era indicates that they were once very plentiful. The irony is that, despite their beautifully colored heads, the puffin's body, with its tight body feathers and short, almost stubby black wing feathers, was not the prize for women's hats on Fifth Avenue. Millinery gunners were more interested in the more flamboyantly long and frilly feathers of herons and egrets, the sharply sculptured wings of terns and gulls, or the spotted wings of owls, shorebirds, and ducks.

But puffins lived on the islands with terns and gulls, and they became valuable in their own way. Their lack of wariness made them easy targets for hungry fishermen, happy to have fresh meat after a winter of salted cod and root crops. To catch the puffins, they draped fishing nets over the boulders at night and then grabbed the entangled puffins as they tried to depart their burrows at dawn. The massive hunting of puffins and other seabirds, from Labrador down the North Atlantic seaboard, was so alarming to John James Audubon that he wrote in 1833, "This war of extermination cannot last much more . . . they must renounce their trade."[5] (This statement is a little ironic from the man who had shot puffins with two double-barreled shotguns to study and paint them.) Of his 1833 visit to a puffin island in Labrador, Audubon wrote, "The poor things seemed not at all aware of the effect of guns, for they fly straight towards us as often as in any other direction." He continued, "I shot for one hour by my watch, always firing at a single bird on wing. How many puffins I killed in that time, I take the liberty of leaving you to guess."[6]

By the 1880s, a bird preservation movement named after Audubon began taking hold as terns and other birds were being wiped out by the hundreds of thousands, including forty thousand birds on Cape Cod in 1883.

The story was just as pitiful in the American Pacific. Albatross and tern hunting had commenced on US atolls on a scale unimaginable today. Theodore Roosevelt was warned in 1904 that hundreds of thousands of albatross and terns were being taken; in response, he began establishing bird refuges where he could. But the countervailing political will on Capitol Hill, in favor of the fashion industry, was summed up in

1913 by Senator James A. Reed of Missouri. When American antiplumage efforts reached the floor of Congress, he mocked: "The swamp herons' afflictions are doubtless solaced by the thought that it is only a miserable, homely creature, of no use on earth except for one feather, and that its departing agonies must be alleviated by the knowledge that that feather will soon go to glorify and adorn my lady's bonnet."[7]

Puffins were among the many victims of this avian holocaust. The bird's exotic beauty evoked neither sentimentality nor shield. Charles Wendell Townsend, the primary contributor on puffins to Arthur Cleveland Bent's *Life Histories of North American Diving Birds,* wrote in 1919 that the puffin "is a curious mixture of the solemn and the comical. Its short stocky form and abbreviated neck, ornamented with a black collar, its serious owl-like face and extraordinarily large and brilliantly colored bill, suggestive of the false nose of a masquerader, its vivid orange red feet and legs all combine to produce such a grotesque effect that one is brought almost to laughter. . . . Besides being grotesque it is singularly confiding or stupid, and it is this, it seems to me, that is leading rapidly but surely to its downfall and final extinction, unless refuges are created and respected where it can breed undisturbed."[8]

The annihilation of puffins on Western and Eastern Egg Rocks and Seal Island and their reduction down to the last pair on Matinicus Rock was met with relative indifference aside from Audubon's alarms in the 1830s and Townsend's pleas nearly ninety years later. Yet despite these persecutions, puffins have never been imperiled sufficiently to rank as federally threatened or endangered species. Today, they are among the fortunate birds that the International Union for the Conservation of Nature lists as of "Least Concern," with a global population deemed "extremely large," though they are state listed as "threatened" in Maine.

Noted puffin authorities Michael Harris and Sarah Wanless estimate that, counting nonbreeding puffins, the global population is about twenty million birds, of which six to eight million are too young to breed. Atlantic puffin nesting colonies stretch in an arc from the Brittany coast of France through the British Isles up to Sweden, Norway, Russia, Iceland, and northern Greenland, across the frigid North Atlantic over to Newfoundland and Labrador and down through the waters of Nova Scotia to midcoast Maine. Iceland is by far the capital of the puffin world, home to three to four million pairs.[9]

Although some colonies are protected as refuges, others are open for limited "harvesting" for meat, which is considered a delicacy. About four to five million puffins nest in the Vestmannaeyjar archipelago off Iceland's south coast, so many that on the island of Heimaey, many pufflings that emerge from their burrows hoping to get their bearings to the open ocean instead plop off cliffs right onto the town streets. For five or six festive weeks in July and August, Heimaey hosts Puffling Nights. Parents suspend bedtimes while children morph into puffling guardian angels. They roam the streets through the night, gathering up thousands of bewildered pufflings by the boxful and tossing them free into the sea come morning.[10]

These are some of the same children and pufflings who, when grown, meet again as predator and prey. Another summer event of the island is its annual puffin hunt, where between sixty and eighty thousand birds have traditionally been taken. The hunters sit high on the grassy slopes and, with a deft wave of a long-handled net, snatch incoming puffins from the air. Each feisty puffin is hauled growling and clawing out of the net into the hunter's hands, where a quick twist of its neck ends its life. The record-holding puffin hunter repeated this act 1,204 times in one day. Puffin hunting in the Faroe Islands, a cluster of eighteen islands that lie northwest of Scotland and halfway between Iceland and Norway, reached a peak of four to five hundred thousand a year in the 1940s, but the number of puffins has crashed in recent years, and by 2010 just 339 birds were captured.[11] The puffin hunting season in the Vest-mannaeyjar archipelago was suspended in 2012 but continues elsewhere on Icelandic islands where puffin nesting is more successful.

On the North American side of the Atlantic, there are about five hundred thousand pairs, with about three hundred thousand pairs residing on three islands in the Witless Bay Ecologic Preserve, near Saint John's, the provincial capital of Newfoundland and Labrador. That may sound like a lot of birds, but this is only a fraction of the number that used to live there. Puffins were probably also abundant on many islands off Nova Scotia in the early nineteenth century, but by 1922 they were largely gone.[12] Puffin numbers have rebounded since about 1980. As of 2013 about four hundred pairs nested at five Nova Scotia colonies with about a hundred pairs in Cape Breton and another hundred pairs at Pearl Island.[13]

Along the north shore of the Gulf of Saint Lawrence, continuous disturbance of colonies and excessive exploitation was so great that by 1906 most islands were nearly devoid of birdlife. Perhaps the most dramatic destruction was at Perroquet Island in Bradore Bay. In the 1840s, Audubon described the puffin population as so immense that "one might have imagined half the puffins in the world had assembled there." Assessments in 1906, 1909, and 1915 revealed that numbers were "probably not one-hundredth part" of what Audubon found.[14]

The most extreme case of persecution was the great auk, the largest of the recent members of the Alcidae. It is easy to look back and say that the assault on this bird should have set off all the alarms necessary to spark a global conservation movement a half century before it actually occurred. On the European side of the North Atlantic, the flightless bird (called "penguin" long before the similar-looking but unrelated birds of the southern hemisphere) was most prominent on islands off Iceland and Great Britain. The hunting took place on the open ocean as well as on land. Scottish fishermen learned that they could lure great auks toward their boats by waving fish at them; when the birds swam close enough, the men knocked them out with a swing of an oar. The last great auks died in 1844 on the Icelandic island of Eldey.[15]

A more chronologically detailed and equally disturbing history of the long-term destruction of great auks is more available from North America. By the time of written records in the 1600s, the colony of great auks on Funk Island, New-foundland, was probably the world's single-biggest remaining population. The island was nicknamed the "island of penguins" by Portuguese sailors around 1520, and within the decade, ships from several European countries were hauling great auks off the island for meat. Jacques Cartier was so taken by Funk Island's bounty during his maiden exploratory voyage to North America in 1534 that he made it his first stop on his second voyage a year later.

It is not clear when the last great auk was taken from Newfoundland, but it was apparently in the mid-1800s, about the same time that the last great auks were killed in Iceland. A final graphic reminder of the Newfoundland extinction came in an 1876 journal entry by naturalist Joel Asaph Allen, who interviewed one Michael Carroll of Bonavista, Newfoundland, who had witnessed the slaughter of great auks. "As the auks could not fly, the fishermen would surround them in small boats and drive them ashore into pounds previously constructed of stones. The birds were then easily killed and their feathers removed by immersing them in scalding water in large kettles set for this purpose. The bodies were used as fuel for boiling the water. This wholesale slaughter, as may well be supposed, soon exterminated these helpless birds, none having been seen there, according to Mr. Carroll, for more than 30 years."[16]

By the 1880s outrage in the United States over the slaughter of birds for feathers was mounting even as feather hunting, too, was gaining momentum. On one front, naturalists and ornithologists, led by Elliott Coues and William Brewster, gathered in New York in 1883 to form the American Ornithologists' Union (AOU). With their eyes set on the legislative arena, AOU members began suggesting model laws that state governments and federal agencies could adopt to begin protecting birds that were not traditional game birds but were being killed for the feather trade.

The AOU's bird protection committee determined in 1886 that on average five million birds were killed annually for fashion, particularly up and down the US coasts. That number may have been conservative.

No bird seemed safe in this era. "Shorebirds, a large and diverse group that includes the plovers and sandpipers, were threatened *en masse*," wrote *Audubon Magazine* field editor Frank Graham Jr. in *The Audubon Ark*. "Most of these birds nest in the far north among the grasses, ice pools and briefly blooming wildflowers of the tundra. On the way to or from the tundra . . . the shorebirds were forced twice each year to run the gauntlet of guns fired by those who were out for sport or food.

"A lighthouse keeper, describing the hunting of 'peeps,' or small sandpipers, on an island off the coast of Maine wrote, 'They form in flocks and sit on the shore. Gunners come here and slaughter them awfully, for it is no trick to fire into a big flock of them and wound a large number. After the gunners have been here, my children bring in many wounded ones, some with broken wings or legs shot off, or eyes shot out, in all shapes. The gunners don't get half they shoot down.'"[17]

At the same time, the appropriately named George Bird Grinnell, editor of *Field and Stream* magazine, decided to try to end the excessive bird killing in America. In 1886 he wrote an editorial proposing "an Association for the protection of wild birds and their eggs, which shall be called the Audubon Society. Its membership is to be free to everyone who is willing to lend a helping hand . . . to prevent so far as is possible, (1) the killing of any wild birds not used for food; (2) the destruction of nests or eggs of any wild bird; and (3) the wearing of feathers as ornaments or trimmings for dress."[18]

Within twelve months, thirty-nine thousand people had signed up for the society, including Oliver Wendell Holmes and Henry Ward Beecher and others with righteous indignation. John Greenleaf Whittier wrote, "I almost could wish that the shooters of the birds, the taxidermists who prepare their skins, and the fashionable wearers of these feathers might share the penalty which was visited upon the Ancient Mariner who shot the Albatross." Charles Dudley Warner, Mark Twain's collaborator on the novel *The Gilded Age,* declared, "A dead bird does not help the appearance of an ugly woman, and a pretty woman needs no such adornment."[19]

Although Grinnell was overwhelmed by the response to his call, this first attempt at a society called Audubon collapsed by the end of 1888. In capitulation, Grinnell said, "Fashion decrees feathers; and feathers it is. The headgear of women is made up in as large a degree as ever before of the various parts of small birds."[20]

Yet the energy behind the failed Audubon Society continued to gain strength. In Boston, Harriet Hemenway, a matron descended from the wealth of the cotton mills, was outraged on hearing reports of heron slaughters. She called a meeting at her home in the heart of the city to form the Massachusetts Audubon Society. That stimulated a burst of state Audubon Society startups. By 1898, there were societies in nearly all the eastern states; in Midwestern states such as Wisconsin, Illinois, and Minnesota; and in Tennessee, Texas, and California. The momentum reached Congress, which in 1900 passed a bill, sponsored by Representative John F. Lacey of Iowa, banning the interstate trafficking of birds killed in states that banned millinery hunting. Some nervous millinery shops began returning feathers to wholesalers when they heard the government was moving in.[21]

A key figure in the campaign was William Dutcher. An insurance man by day, he was such an avid conservationist in his spare time that he chaired the American Ornithologists' Union's committee on bird protection in 1896 and 1897. On one occasion he tipped off federal agents to a twenty-six-thousand-gull-skin stash in Baltimore.

After the Lacey Act was passed in 1900, Dutcher began lobbying state legislatures for stronger laws to halt the hunting of threatened birds—period. At the same time, he took matters into even more personal hands by teaming up with wildlife painter and naturalist Abbott H. Thayer to keep hunters away from sensitive bird areas. Thayer's art was known for its shrewd shadings and concealed colorations that some credit in the evolution of military camouflage. Thayer gave Dutcher the first fourteen hundred dollars toward a war chest to hire wardens to protect birds at the point of a gun.[22]

Maine was the first beneficiary.

In March 1900, Dutcher had received a disappointing letter about the state of seabirds on Matinicus Rock from ornithologist Manley Hardy. Hardy reported: "The keeper of the light, a Mr. Grant, did all in his power to protect them, but men would come and lay in boats and put out wounded gulls for decoys and shoot in spite of all he could do." Now armed with cash, Dutcher appealed to the lightkeeper, William Grant, to step up his protective efforts with a $25 bonus. That did not sound like a lot of money for a whole season, but Grant wrote back to Dutcher to say that he would do his best to protect the birds without compensation.[23]

But before he was named the first Audubon warden, Grant died suddenly. He was replaced by James Hall, who was described as "the finest shot in these parts." Hall's reputation was clearly a deterrent to the gunners. In the summer of 1901, Hall reported that terns and guillemots had a good nesting year. He also mentioned that puffins "raised young during the season." It was the first and only known prior effort to preserve puffins in Maine. Hall and his rifle may have very well have saved the last pair of puffins in Maine.[24]

By 1904, Maine would have ten wardens, the most of any state. Despite the low pay, Dutcher was adamant that they do their job. His tone was set in a stinging admonishment in July 1901 to Captain George D. Pottle, the bird warden assigned to protect birds in Muscongus Bay: "A member of our society has just returned from a visit to the Eastern and Western Egg Rocks, and he writes me that the actual conditions there are deplorable. He states that the islands have been visited and from all the evidence he could obtain, have been systematically robbed. He also states that he found places where petrels have been dug out.

"I would like to know what you have to say about the condition of these islands, and what you have done to protect them as you agreed. I cannot pay you for work that you do not do, nor do I care to pay if the birds are not protected. I expected that you would keep people from taking eggs, which is contrary to law, just as much as I expected you to prevent people from shooting the birds.

"If you were aware that people were disturbing the birds breeding there, it was your duty to inform me so that I could take steps to have them prosecuted under law."[25]

These efforts led to the federal government's establishment of the first national wildlife refuges, and Dutcher would go on to become the first president of the National Association of Audubon Societies in 1905. But those efforts, no matter how valiant (Guy Bradley, a wildlife warden, was killed in 1905 when he confronted egret hunters in the Florida Everglades), could not lure back certain birds, such as puffins, to places where they had been eliminated. The ensuing decades offered no new road maps for anything more ambitious. In fact, by the time I arrived at Hog Island, the prevailing notion of seabird conservation was nothing more than passive, hands-off management. Seabird conservation was underfunded by both federal and state governments, which barely had the resources to count birds and place small signs stating that the islands were closed to landing in summer. Even the ownership of some important nesting islands such as Eastern Egg Rock was unknown.

Although this situation resulted in part from the chronic

lack of funds for seabird conservation, it was also because of a do-nothing, nature-will-take-care-of-itself notion that ran deep even among serious ornithologists. Leaving seabirds alone to recolonize at their own pace did in fact work for gulls, eiders, cormorants, and guillemots. But other species, such as puffins, murres, gannets, and terns, remained absent from their historic Maine nesting islands.

Majestic eagles and peregrines had human advocacy on their side to reintroduce them in the contiguous forty-eight states, but there were no active programs in place to bring back lesser-seen, lost seabird species. For these birds, the general thinking was "let nature take its course." Even as researchers were beginning to transplant raptor eggs and chicks, the ethos in the seabird conservation world was that it was wrong to "play God" and that a new "balance of nature" would set in—consisting of those species that could survive around humans. This thinking did not bode well for puffins.

This passive approach made no sense to me: it was humans who, through excessive hunting, caused the birds to disappear. I guessed that the main reason puffins, terns, and other seabirds had not returned to their historic homes once they were safe from hunting was the existence of the large gull populations that were responding to the mountains of garbage deposited in open landfills and the waste from local fisheries, especially lobstering. Although terns repopulated some key nesting islands following the end of market hunting, gulls apparently wiped out the last tern populations on the Egg Rocks in the 1930s. By the early 1970s the few remaining tern colonies in Maine faced chronic attack from predators because gulls had displaced them from safer islands farther offshore, forcing the terns to nest closer to the mainland, where they had to contend with a long list of mammals and avian predators such as crows, great horned owls, and black-crowned night-herons.

My original idea was to collect puffin eggs from an existing colony and move them to one of the Egg Rocks, where I would swap eggs with black guillemots, an alcid cousin. Whereas puffins have only one chick at a time, guillemots usually have two. Regardless, I hoped that the guillemots would incubate and rear the puffin chicks to fledging age. I also hoped that the young puffins would remember the release site and come to think of the Maine coast as home. This idea of "cross-fostering" was then being considered for the establishment of a new population of endangered whooping cranes through enlisting the more common sandhill cranes of New Mexico as "avian midwives."

I reasoned that if successful, swapping out eggs would shortcut the more complicated approach of translocating puffin chicks and hand-rearing them to fledging age. Yet there were many unknowns associated with cross-fostering. Would the parents accept puffin eggs (guillemot and puffin eggs look very different)? Would puffin chicks thrive on a meal of rock eels rather than herring and other more typical puffin forage fish? Most worrisome was the concern that puffin chicks might become imprinted on guillemots and be confused later in life when they went to select mates (this eventually was the case with the cross-fostered whooping cranes).

On further consideration, I concluded that, although cross-fostering might be worth these risks for highly endangered species in which every egg and chick was critical, fortunately puffins were far from endangered, and moving chicks made more sense. During the summer of 1971, I continued to read about puffins and to reach out to anyone who I thought might offer some wisdom.

It was soon apparent that all roads were leading to William Holland Drury Jr., one of New England's leading seabird experts and research director at the Massachusetts Audubon Society. I wrote a letter and a draft proposal in August 1971 sketching out my plan for a puffin comeback and was thrilled to receive not only a quick response but an invitation to meet in person at his office at Drumlin Farm Wildlife Sanctuary in Lincoln, Massachusetts.

I found the esteemed Dr. Drury tucked away in a back office surrounded by stacks of papers and books in a historic building on the Mass Audubon property. Tall, lanky, and mostly bald, he welcomed me warmly, and we had an excellent, refreshingly upbeat talk about Maine seabird islands. Years earlier, he had conducted gull censuses on the Muscongus Bay Egg Rocks and was familiar with their history as former puffin nesting grounds.

I did not realize it then, but by being in Drury's office, I had stepped into the inner sanctum of ecology. Before becoming research director at Mass Audubon, he had been a lecturer for two decades at Harvard, where he became a close colleague of Ernst Mayr, often acclaimed as the greatest evolutionary biologist of the twentieth century.

Mayr was reared in Bavaria by parents who loved hiking in search of flowers, fossils, and mushrooms, and by the age of ten young Ernst could identify all the local birds. Years later he would assert that his early exposure to the outdoors was critical to his development. He believed that few people pick up an interest in nature after childhood, and those who do are most likely to focus on just one area, such as bird-watching, without developing a deep understanding of the surrounding ecosystem.

Mayr eventually came to the United States to work as an

ornithologist at the American Museum of Natural History in
New York and went bird-watching with Irv Kassoy's Bronx
County Bird Club. Where the Bronx boys showed him American birds, Mayr introduced them to scientific ornithology,
influencing how Kassoy studied his barn owls in painstaking
detail in the haunted mansion in Pelham Bay. Mayr moved to
Harvard in 1953, a year after Drury began teaching there.

William Drury was the son of artists who encouraged
him to roam the beaches and woods near his native Middle-
town, Rhode Island. In a childhood parallel to Mayr's, he was
excused from organized sports in his private school in lieu of
a personal program in which he walked a six-to-eight-mile
birding route. He also took up drawing the birds he saw. "Drawing required that I watch individuals closely in order to recognize them by their personality and put that down on paper:
their proportions, how they moved, how they perched, how
they scolded, how they expressed alarm, how they flew," Drury
later wrote. From direct observation in nature, he worked to
inform himself about how birds interact with nature.[26]

In his foreword to Drury's memoir *Chance and Change,*
Mayr observed that "Drury was not simply a describer who
conscientiously recorded all he saw, but whenever he looked at
the natural scene, he saw puzzling irregularities, unanswered
questions, contradiction, and what others call exceptions."[27]

Drury was deeply influenced by Mayr's scientific doggedness as Mayr pursued yet undiscovered elements of natural
selection. He was once quoted as calling Mayr a "breath of
fresh air," someone who inspired him to challenge the notion
that "nature's norm is balance." Instead, said Drury, the sum of
what he observed in the field was "neither pervasive order nor
chaos, but comfortable disorder."[28]

Drury's view of challenging the prevailing thinking about

the way the world works is summarized nicely by this statement: "When your views on the world and your intellect are
being challenged and you begin to feel uncomfortable because
of a contradiction you've detected that is threatening your current model of the world or some aspect of it, pay attention. You
are about to learn something."[29]

His combination of doggedness and challenging the order
of the day was already receiving wide recognition by 1956, a
time when the conservation movement was entering public
consciousness (Rachel Carson had published *The Sea Around
Us* and *The Edge of the Sea* by then). At this time, Massachusetts Audubon launched a science department and made Drury
its first director. In this new role, he pioneered studies of plovers, gulls, and warblers.

In October 1960, after an Eastern Airlines Electra plane
crash at Boston's Logan Airport killed sixty-two people when
a flock of starlings was sucked into the plane's engines, Drury
was called by the Federal Aviation Administration to help find
out how to reduce risks from bird strikes. In Civil Aeronautics
Board hearings three months later, Drury testified that the
starlings were likely "clumping." As paraphrased by Michael
Kalafatas in his book *Bird Strike* about the incident, Drury explained that clumping was "an instinctive reaction to prevent
a predator from singling out one bird, which is what a peregrine falcon . . . does when it plummets close to 200 miles per
hour to pick off a lone starling for lunch. Starlings usually fly
in a strung out pattern, but when frightened, they clump into
a mass, dense and deep."[30]

Drury was able eventually to convince officials why star-
lings and much larger birds that also can cause crashes were at
Logan in the first place. He would say years later, as recalled by
Frank Graham Jr., that if the people who designed Logan had

set out to make a bird sanctuary instead, they couldn't have done a better job. Drury noted that the airport was surrounded by city dumps and sewer outfalls where gulls could feed. The airport even had a dump of its own, and the area was dotted with freshwater ponds and salt marshes where gulls could loaf. Finally, it had long, isolated runways stretching into the harbor where the gulls could roost undisturbed at night. It was no wonder that gulls and starlings abounded at the site.

In 1962, Drury's expertise earned him a position on John F. Kennedy's Scientific Advisory Committee that was investigating the pervasiveness of pesticides in the explosive wake of the publication of Carson's *Silent Spring*. President Kennedy publicly offered sympathy for Carson's concerns of mass contamination of the environment, but his administration was also being hammered on the inside by the chemical industry and its apologists within the Agriculture and Commerce Departments. As Zuoyue Wang recounts in the book *In Sputnik's Shadow: The President's Scientific Advisory Committee and Cold War America,* the industry wanted pesticides considered innocent until proven guilty.

When not playing a part in this seminal moment of environmental awakening, Drury set himself to a general task of changing the very definition of *environmentalism*. Some of his concepts were as simple as debunking the herring gull's reputation as a "pest" when it was only doing what it was equipped to do—that is, taking advantage of human trash. Others were more complex, such as sounding an alarm on what he saw was a growing chasm between humans and the natural environment. In his concern, he challenged not just prodevelopment forces but also many of his naturalist allies. "Some consider their responsibilities done when they have established wilderness where the affluent can enjoy their safaris or canoe trips," he said.[51]

Drury was worried that both environmentalists and developers in their own way tend to separate humans from nature. This attitude implies that humans and their habitats are not natural because they have consistently created imbalances. He believed that environmentalists continually assert that humans and their technological society have destroyed nature. They argue that before humans appeared on the scene, all was peaceful and harmonious. He held that "this is a distorted, unsubstantiated view" and that "the dichotomy between human-influenced systems and 'natural' systems is not realistic or helpful" because "it leads to an unjustified pessimism among environmentalists." The idea that humans are separate from nature, he observed, "relieves us of any responsibility for nature."[32]

The separatist view leads people to believe it is possible to protect nature by creating nature preserves and that these can function without human intervention. Drury was among the first to see that human influence was affecting nature everywhere. Regarding the need for human actions to save wildlife, he wrote, "As to the philosophical issues of nature's order and 'playing God,' we now know that laissez-faire ecology, like laissez-faire economics, doesn't lead to balanced systems, it leads to monopolies. Unless we believe that there is a natural order established at the Creation, we should acknowledge that when we won't play God, someone will." I could see Bill Drury's point at the Egg Rocks. Without directed action to help the displaced species such as puffins and terns, the island was occupied by a monopoly of adaptive gulls that were thriving from the largesse of our throwaway lifestyle. The gulls were playing "God."[33]

I found his thinking very consistent with my own emerging ideas about stewardship. I didn't know it yet, but when I left my meeting with Drury, I had not only a sharper focus on a plan but a new ally for moving through a long maze of obstacles.

My plan for bringing the puffins back to the Egg Rocks was taking shape. I would need to find a suitable source colony and then move either eggs or chicks to Maine and somehow rear and release them. I hoped the pufflings would learn that Maine was home rather than their natal colony. This made sense to me because of the puffin's life history. Puffin chicks usually hatch in June in a dark burrow as much as eight feet deep. They are fed small fish by their parents until they are about six weeks old, at which time they lose interest in eating, become very restless, and eventually scramble off into the ocean. With wings usually still too small to fly, the puffin fledglings paddle toward the horizon in the dark—their best strategy for avoiding predatory gulls, which usually sleep at night. In this way, puffins begin their life at sea alone with a built-in set of instincts to help them dive and pursue small fish and other small marine creatures. By dawn, the fledglings are already out of sight of home, drawn by ancient instinct toward the vast North Atlantic.

At the time, little was known about the behavior of puffins in the years before their first nesting. For example, it was unclear how many years puffins stayed at sea before deciding to come back to their home islands. It was also unclear if they visited different islands for the next two or three years or made a beeline to the place of their hatching. Later we learned from our own studies that a few puffins begin nesting when they are four years old, but most are five or more when they first begin nesting. We also learned that there is great variability in the tendency to return to the homeland. Some puffins are very faithful to their homeland, while others visit several puffin colonies before nesting for the first time.[34]

Several features of the puffin's life history gave me the sense that puffin chicks could be moved as chicks from one colony and released at another—with a good chance that they would return to the release site. First, from the day they hatch, puffin chicks eat whole fish completely on their own after the food is delivered to the burrows by their parents. Second, fledglings are not escorted to sea by their parents and they naturally develop feeding, survival, and migration habits on their own. All I had to do, I hoped, was fledge enough healthy chicks and then wait for success.

All of this was clearly a huge gamble because no one had ever translocated puffin chicks before, and the only previous translocation projects with seabirds had both failed. Projects that attempted to move fledgling-age Laysan albatross from Midway Island to new nesting islands in the Pacific and short-tailed shearwaters in Australia to new homes both moved fledglings from large colonies and released them at new sites with hopes that the chicks would return to the release site to nest. In both projects most of the young eventually returned to their natal colonies rather than the release site. These were troubling outcomes because moving fledglings would be much easier than moving eggs or downy chicks.[35]

I reasoned that if translocation of chicks was going to work for puffins, I would need to move them as young as possible. And since brooding by the parents was thought to end at about ten days of age, I chose ten-to-twelve-day-old chicks as my preferred age for the translocation. One thing was certain, however—we would need to serve as surrogate parents for the

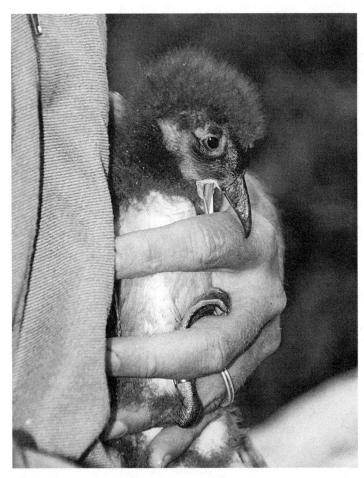

My plan to revive the Eastern Egg Rock colony focused on moving puffin chicks that were about ten days old. My hope was that they had not yet learned the location of their natal home and that they would eventually return to the release site at Egg Rock. We banded the chicks near fledging age (above). Photo by Derrick Z. Jackson

chicks. We would need to build burrows and remain alert night and day to chase off gulls and other avian predators and provide high-quality meals, which in this case meant adding vitamin supplements to thawed fish of just the right size. This plan seemed so clear to me that I could already imagine myself sitting at night by the edge of the sea, waiting like an expectant father, to witness the pufflings making their epic journey to the sea.

It was also apparent from the beginning that I would have to displace the great black-backed gulls that dominated the Egg Rocks. I began floating my nascent proposal to Maine state wildlife officials and waited for a response. Eventually, I was told they might approve my plan to bring the puffins back—but they needed a year or two to consider my full proposal because of the need to manage gulls. Perhaps they thought I would drift on to some other project with sufficient delays and avoid any controversy around gull control. But they had no idea that "stick-to-it-ism" was already my hallmark. That term was first applied to my developing bird handling by my seventh-grade Bexley science teacher Zale Thorleau, who applauded my success in building a homemade chicken incubator (after many failed attempts) for a science fair. I had no way of knowing at this early stage of Project Puffin how important my inclination for persistence would become.

It would have been easiest to obtain puffin chicks on Matinicus Rock thirty-five miles to the east, since the Audubon Society had a long history managing this outermost bird island. Also, the chicks would not have to cross an international border. But though lighthouse keepers and their guns preserved the existence of puffins from human hunters, beginning with sharpshooter James Hall, there were still few puffins, and it was the only island south of the Canadian border with puffins. Matinicus Rock was home to only about seventy puffin pairs in the early 1970s, and the possible impact of taking chicks from such a small population ruled that island out from the beginning.

As a second-best location, Bill Drury suggested Machias Seal Island, located at the US-Canadian border, about 105 miles east of Eastern Egg Rock. After Matinicus Rock, it was the nearest puffin colony to the Egg Rocks. He thought that the thousand-plus pairs of puffins there could provide enough chicks without affecting the colony. So on Drury's advice, I wrote to David N. Nettleship of the Canadian Wildlife Service (CWS). As its Eastern Region seabird research scientist, he was also a leading expert on puffins and gull predation on seabirds.[36] His position with the CWS also made him the gatekeeper for the permits that I would require.

With a hopeful cover letter, I sent Nettleship my draft proposal for collecting puffin eggs and chicks on Machias Seal Island for the purpose of reestablishing a puffin colony on the Egg Rocks of Maine.

But his response, dated February 18, 1972, seemed to slam the gate in my face even more firmly than Ralph Palmer's naysaying letter. He agreed that it was a great goal to reestablish the lost puffin colonies of Maine and that I had put a lot of thought and careful planning into my proposal. However, he believed that Maine's location at the southernmost end of the puffin's North American range made it a "peripheral" breeding ground. He continued to explain that he had documented decades of decline on both sides of the Atlantic, including the north shore of the Gulf of Saint Lawrence, southern England, and Norway. He cited many reasons for the declines, including climatic changes, changes in food availability, oil spills, chemical poisoning, and other pollutants. He concluded that when such declines happen, the remaining birds retreat toward the center of their ranges, resulting in extinctions of the peripheral colonies. Nettleship bluntly concluded that this behavior would make any attempt to reintroduce puffins to Maine futile because my plan had too many possible sources of error.

Interestingly, he ended the letter saying something I completely agreed with. Nettleship wrote that the best way to ultimately protect seabirds was to govern human impact on their marine environments through laws and public awareness campaigns. Ironically, this very conclusion inspired me because I hoped that a restored colony near the Hog Island Audubon Camp would lead to greater awareness for seabird conservation and a following that would support policies to protect seabirds, forage fish, and ocean habitats.

Like the Maine state officials, David Nettleship probably thought he would never hear from me again. I put down his reply, feeling as much sting as I did from the letter from Ralph Palmer.

3

My Judge and Drury

I learned two and a half weeks later that Nettleship's rejection of my request for puffin eggs and chicks from Machias Seal Island had been seconded at the highest levels of the Canadian Wildlife Service. A letter from W. R. Miller, writing on behalf of both J. E. Bryant, Eastern Region director for the CWS, and F. Graham Cooch, director of the entire CWS, advised me that in their opinion it would be irresponsible to authorize the collection of eggs or chicks. They explained that their responsibility was to protect the island and that my proposal would deliberately deplete the colony—and this was "contrary to their objectives."

With such sobering news, I wrote Bill Drury, hoping that he could recommend next steps. I recalled reading a published letter that he had written in support of Tom Cade, pioneer of the Cornell peregrine restoration program, when Cade was under attack for releasing falcons from diverse locations into the eastern United States. Drury understood the need to take action on behalf of seabirds and was frustrated by those who critically called management "playing God."

I had hopes that Drury might defend my ideas because

of his unconventional rejection of passive management and his belief, captured in his memoir *Chance and Change*, that the effects of human activities should provide opportunities to observe how nature responds to modification and change.[1]

Drury believed that a widely dispersed population had the greatest chance of survival, that reduced ranges greatly increased the likelihood of extinctions. This view supported my belief that puffins nesting at just two colonies along the Maine and New Brunswick coasts were unacceptably vulnerable should predators or an oil spill affect their limited nesting areas.

In my defense, on March 28, 1972, Drury sent off a long letter to David Nettleship explaining why he thought the puffin project was worthwhile. He explained that the puffin colonies on both Machias Seal Island and Matinicus Rock were recovering from the nineteenth-century hunting era and that owing to the small number of puffins in the Gulf of Maine, a "seed colony" was necessary to encourage young puffins to pioneer new colonies. He concluded that puffins needed this help and that they would respond favorably to a translocation plan.

Bill concluded his letter to David saying that he hoped we could all work together to create a program with maximum promise of success. I especially appreciated his appeal "not to close the door on a young man's immature plans." He wisely observed that learning and experience are part of a graduate study program and this was something that both he and Nettleship had previously experienced.

Two weeks later, Nettleship responded to Drury, telling him that he was encouraged by this assessment of the Maine-Fundy puffin population. But Nettleship was still not convinced that small increases indicated a "rosy future" for puffins off the Maine coast. Although he had dismissed my first round of

ideas, the discussion with Drury had clearly gotten him thinking about how to proceed. Nettleship began to offer his own
suggestions and mentioned that he had personally enjoyed
success hand-rearing chicks. He suggested:

1. Collect puffin eggs soon after egg-laying (to allow
   adults to lay another egg);
2. Hatch the eggs in an incubator;
3. Rear chicks in captivity until they reach fledging
   condition; and
4. Transfer them to natural or artificially dug burrows
   at extinct breeding colonies where they will be left to
   fledge "naturally."

Nettleship added that if some form of "imprinting" did
not occur at the time of fledging, the period spent in captivity
should not alter the outcome, especially if chicks were reared
under conditions somewhat resembling their natural ones. I
was happy to hear about his previous experience rearing puffins in captivity on a diet of thawed frozen smelt, supplemented
by multivitamin pills.

And I was elated that, with Drury's seasoned help, Nettleship was beginning to think about reintroduction. In one
correspondence with Drury, cc'd to me, Nettleship still referred
to my proposal as poorly designed and researched. But I was
so thrilled by the turn of events that this comment did not
matter as much as his shift in thinking.

Still, Nettleship was concerned that the work with puffins was
part of my graduate program and the outcome was such a
huge gamble. But after I clarified that the project was only a
minor part of my graduate plan, he became much more coop-

Black guillemots and puffins are both in the family Alcidae. I hoped to learn about raising puffins from black guillemot chicks, but my nascent experience with guillemots ended in disaster. Photo by Derrick Z. Jackson

erative and willing to work together; we arranged to meet in Ottawa in August 1972 to discuss next steps.

With encouragement that I might see the first puffin chicks the following summer, I used the summer of 1972 to gain some experience by rearing black guillemots. I thought that guillemots would help me develop puffin chick-rearing methods because they are in the same family (Alcidae) and have a similar life history. Both species are rock crevice nesters that eat fish as chicks and have a fledging strategy where the chick leaves the island solo without its parents.

Fortuitously, Bill Drury published a very thorough two-part article that gave credence to the causes of New England

seabird demise the same year in the journal *Bird Banding*. In
"Population Changes in New England Seabirds," he reported
that, like puffins, guillemots had been "essentially eliminated"
by the millinery trade and the hunting and egging sprees of
the late 1800s. By 1903, Arthur Norton of the Portland Museum of Natural History estimated, there were perhaps 75 pairs
of guillemots left on only 14 of Maine's 4,613 islands. But somehow they were able to come back on their own in a way the
puffins could not. By 1931, Norton counted 300 guillemot pairs
on 24 islands. Four decades later, in a three-year survey that
ran from 1970 to 1972, Drury estimated 3,400 pairs in coastal
Maine colonies.[2]

With permits in hand, I collected six guillemot chicks
from nearby Franklin Island, where they were nesting in the
rubble of a demolished foundation of an old lighthouse keeper's home. I started out rearing them in cardboard boxes lined
with newspaper on the porch of the Binnacle, an early 1900s
Hog Island cottage that I shared with other camp instructors.
This was such a foreign environment for the chicks that I at
least wanted to offer them rock eels, their natural food, which
often hid under rocks at low tide. That effort quickly unraveled
into a sobering experience when I couldn't find enough of the
ribbon-shaped fish to begin to meet the demands of the chicks.

I found so few rock eels that I was forced to use frozen
fish. Because only large herring were readily available, I needed
to cut these to make them a manageable size for the little chicks.
But this proved a poor alternative: oil dribbled out of the sec-
tioned fish and stained the birds' breasts. It was also impossible to keep the chicks from fouling their feathers in their own
excrement.

When the first guillemot reached fledging age and sported
its full set of juvenile feathers, I released it into the ocean, but

as it swam away, it sank lower and lower, until only its head was above water. Something was dreadfully wrong. The sad little bird lacked waterproofing! I tried to retrieve it, but it had paddled off into deep water and the fog soon engulfed it. Watching my first fledgling sink into the Atlantic was a tragic and humbling moment.

Shocked by this disaster, I tested the other fledglings' waterproofing by floating them in an antique claw-footed bathtub in the Binnacle. As my sympathetic cabin mates looked on, I saw that the remaining five guillemots all suffered similar waterproofing deficiencies. After checking with a few aviculturists, I learned that guillemots were especially prone to this problem, but no one could offer a practical remedy. That summer I returned to Franklin Island as part of my Hog Island teaching and marveled at the excellent condition of wild-reared guillemot chicks. I saw, with envy, how the wild-reared birds had immaculate plumages when reared by parent guillemots in simple rock cracks. From all of this, I learned an important rule about chick rearing: keep it simple and try to replicate the natural rearing conditions.

By mid-August 1972, I had finished my summer responsibilities at Hog Island and unpacked my belongings from Yellow Springs, Ohio, in upstate New York to begin life as a twenty-six-year-old graduate student at Cornell. I had arranged for a country house near Trumansburg, an idyllic spot on a quiet road overlooking Cayuga Lake about thirty minutes from campus. I should have been completely immersed in my graduate studies, but my mind was racing with excitement about the meeting I had scheduled with David Nettleship in Ottawa a few days later.

I was surprised to find that David was not much older than me. He had a head of curly hair (unlike me) and a huge

beard that I imagined helped keep him warm studying northern seabirds. He talked of his respect for Bill Drury and his
research with puffins. I was especially keen to hear about his
familiarity with keeping puffin chicks in his lab—experience
that seemed especially helpful. I could see his skepticism fading as we discussed details about transplanting some chicks as
a pilot study in 1973. Equally important, I could also see that
the ideas David suggested would likely receive approval from
the Canadian wildlife officials.

Five weeks later, I sent a revised proposal from my graduate student office at the Cornell Laboratory of Ornithology.
Through the winter I further refined the proposal to Nettleship's satisfaction and immersed myself at Cornell. I also submitted a proposal to National Audubon to help defray some
of the expenses of the first season and was thrilled to receive
news early in 1973 that my "external Audubon grant" was approved. I was to receive enough, the grand sum of $2,700, to
purchase an outboard motorboat and a landing skiff and to
hire an assistant for the coming summer.

With funding in place, I was becoming concerned about
the next steps and whether permission would be granted for
the project to begin in 1973. In February, I gave Nettleship a
call for an update. I was still hoping to convince him to give
me access to Machias Seal Island, but he was way ahead of me.
I learned from that call that I would in fact receive permission
to collect six puffin chicks—not from Machias Seal but from
Great Island in Witless Bay, Newfoundland—puffin mecca of
North America, where about 160,000 puffin pairs nested! I was
so excited, I could hardly breathe.

I soon realized how many details needed to come together before mid-July. I needed permits and had no idea how
many or from whom they would come. How was I going to get

to Newfoundland, land on Great Island, and rear the chicks? There were dozens of other unanswered questions. I began hustling on all fronts while juggling my academic commitments at Cornell.

I also called Kathy Blanchard to share the good news. Blanchard had been hired as a student assistant to work in the summer of 1972 at the camp store on Hog Island, which was managed by Dur's wife, Peggy. Like all student assistants, Kathy was asked to conduct a study of her own design and have a mentor among the instructors suggest ideas and offer help as needed. In this role, I had asked her to photograph and keep a journal of the progress of a hermit thrush nest. She was so meticulous in her study that I then asked her to come with me to Popham Beach on the mainland and help me photograph least terns.

Later in the summer I told her about my hope to bring back the puffins the next year. We worked in a trip to Machias Seal Island, where she first saw the puffins herself. And I was beginning to enjoy her wide-eyed enthusiasm for everything in nature in a way that began to blur the edge of my professional mentorship. We had discussed the possibility of her returning to Maine if the project received a green light, since I would be able to hire a summer assistant to look after the chicks. Now it looked like it would happen.

To be on the safe side with the Canadian wildlife officials, I sent another copy of my scientific capture permit request, with my outline for the project. Of course, this all assumed that I would be able to keep these first six chicks alive in the first place. Having gotten this far, it hit me anew how fragile my dream remained, hinging on the fate of fuzzy seabird chicks that no one had ever tried to move for the purpose of restoring a lost colony. And I was rightfully troubled by the unfortunate demise of my black guillemot chicks, which were all eventually released, but in less than ideal condition. My project had even greater stakes because of the previous failures with Laysan albatross and short-tailed shearwaters. If my plan also failed, that would provide further evidence that transplanted seabird chicks would not return to the release site and future attempts to restore seabird colonies using this method would also be doomed.

I had to prove the naysayers wrong.

4

A Suitcase of Puffins

The animal on everybody's mind in June 1973 was Secretariat, the first Triple Crown–winning horse in a quarter century. For much of the rest of the world, it was a summer of tumult. Watergate was exploding in President Nixon's face, and the environment was often at the top of the news. That spring, three years after the first Earth Day, the Supreme Court upheld by four-to-four with one abstention a lower court ruling that barred the Environmental Protection Agency from approving any state air pollution that lowered standards in the development-crazy and mining-dependent Southwest and West. A *Boston Globe* editorial praised the ruling: "So far, neither the regulations nor the deadlines have sufficiently served to protect the atmosphere around us."

Secretariat and even the scrapping over clean air were the farthest thing from my mind. There were still many details to sort out regarding translocating puffins. Four summers after conceiving of the idea of reestablishing a puffin colony, I was on the verge of launching the project and preparing for my first visit to Great Island in Witless Bay, Newfoundland, with

Kathy Blanchard as my first research assistant. She hailed from middle-class Farmington, Connecticut, but her lineage went back to a great-grandfather who was a fisherman, seal hunter, and captain of a schooner that transported people, goods, and coal for the lighthouses between Cape Breton, Nova Scotia, and southwest Newfoundland. He sailed his ship into his seventies and died seven days short of one hundred. Kathy's father and grandfather were also from Newfoundland, with her grandfather serving as the manager of the first fish hatchery in Connecticut. Kathy's family hungered for nature and frequently took her on hikes and adventures in the woods.

This led to a critical moment where her high school guidance counselor told her that there were not many job options for a girl who loved the environment. She was told she could be a forest ranger but that she would be sitting alone in a tower. Her grades were good enough to get into the University of Pennsylvania, where she was far from happy the first year. She tried to apply for summer environmental jobs but got nowhere until a cousin told her about Audubon's summer camps. Kathy applied, writing, "I want this more than anything in my life."

Her desire to join me for the adventure to Newfoundland did not surprise her hiking parents, but they were worried about this business of reaching into burrows along high cliffs, and who was this Steve Kress? They were so worried that Peggy Morton had to reassure them in an exchange of letters that all would be okay. The parents would approve only if a friend of Kathy's—Susan Kains (a fellow intern at the Aullwood Audubon Center)—could accompany her. Ira Gavrin, a former Audubon Camper with a rugged jeep, provided transportation for the expedition.

Nettleship had agreed for us to make two trips to Newfoundland in 1973. Our first trip was a mid-June scouting trip

to see the puffin habitat and to collect temperature measure
ments within puffin burrows to help us replicate similar burrow climates in Maine. After arriving in East Bauline, we soon
found John Reddick, David Nettleship's primary transport to
Great Island. There were fewer than twenty small houses and
just one dock, so everyone knew John. He and his family
had fished for cod and salmon in Witless Bay for decades. He
was a large man with a grand smile, and he gave us a warm
welcome.

We camped near the town landing for the night, then
met John and his brother Bernie at dawn the next morning for
the trip to Great Island. We started encountering puffins nearly
as soon as we left the dock, and the numbers increased as we
approached the island. Some birds were so heavy with meals
of capelin that they could not become airborne even though
they tried running on the water. Soon, Great Island loomed
ahead surrounded by a few Greenland-born icebergs. From a
distance, it appeared to be covered in snow and ice, but as we
approached, we could see that the cliffs were caked in guano.

The slopes were largely covered with dark green grass
and the interior with stunted spruce. John pulled his dory into
a grotto where he secured a rope across two steep cliffs that
Bernie deftly snagged and held with his boat hook. The dory
rose and fell six to eight feet with the sea surges. In between
waves, John shouted for us to jump, and one by one we hopped
out onto the slimy rock, trading the thumping sound of his
antique one-lung motor for the tumult of trumpeting murres
and chanting kittiwakes.

Puffin burrows were everywhere: on the grassy slopes,
among the tangled roots of the spruce forest, and in steep cliff
habitats. In flat regions the grass was lush and green, benefit-
ting from the guano of puffins and more than a million pairs

of Leach's storm-petrels that honeycombed the soil. In some places, the tunneling and excrement from puffins had killed the vegetation, exposing the soil to erosion. Nearly everywhere, puffins stood at burrow entrances or flew in huge ellipses out to sea and back to the island.

These burrow examinations were one of our first reality checks. Although there were about 160,000 puffin pairs nesting here, obtaining large numbers of similarly aged chicks would not be easy. We could only measure temperature in the burrow nest chambers on the steepest, slipperiest slopes, where a missed step would send us tumbling down the cliffs, because most of those on flat habitat were far too deep, sometimes six feet or more.

Puffins dig their burrows using the beak as a pickax and the webbed, clawed feet as shovels. But this takes a lot of work, and a good, secure burrow represents a site worth returning to year after year. Most of the burrows that we could reach were testimony to the puffin's tunneling talent, usually bending at least 90 degrees, some to the left, others to the right. I wondered if this was the equivalent of right- versus left-handedness or simply working around obstructions such as rocks. Clearly, the burrows were an evolutionary response to nosy gulls intent on obtaining a meal of puffin egg or chick. If we encountered an adult in a burrow, it would either turn its back to us, blocking access to the egg, or attack our probing fingers while making guttural, moaning growls.

But the burrows provided another perk for puffins. Our temperature readings demonstrated little variation from one burrow chamber to the next, by day and night. They were always about 55 degrees Fahrenheit.

Early on the third day, we heard the distinctive thumping sound of the motor on John Reddick's dory as it pulled into the landing cove to pick us up for our return to the mainland and the long drive, mostly over rutted dirt roads, to Port aux Basques, where we were to catch the ferry to Nova Scotia. We would return in just four weeks, and there was much to prepare back on Hog Island for our first six puffin chicks.

In the meantime, I still had a lot to learn from guillemots to help me be a better puffin foster parent. I hoped for better success than my first sorry attempt the previous summer.

We collected thirteen guillemot eggs at Franklin Island and put the eggs into a special T-shirt on which we stitched pockets for each egg to bring them back to Hog Island. I wore the shirt under a bulky down jacket to keep the eggs warm. Once back at the camp, we put the eggs into a homemade incubator similar to one I had used in junior high school to hatch chickens but soon realized that it was not holding a steady temperature. Fortunately, we were able to borrow a clinical incubator (intended for bacteria testing in Petri dishes) from a nearby medical office to use while I tinkered with my homebuilt incubator.

For the next three days, Kathy and I took shifts on a round-the-clock schedule, giving the eggs a quarter turn every four hours. As far as we knew, no one had ever successfully incubated a guillemot egg, especially in a bacteria incubator that would not budge past the mid-90-degree Fahrenheit mark. I called the poultry department at the University of Maine to ask how much we should be turning the eggs. They admitted to knowing nothing about seabird eggs, but suggested turning them 180 degrees every eight hours. We gathered this information early in the incubation period and quickly applied the recommendations.

After three days of fussing with my homebuilt incubator,

I was getting the temperature between 100 and 104 degrees, the norm for chicken eggs. Then near midnight on the third night of incubation, Kathy thought she heard a squeak from one of the guillemot eggs. On the fourth morning, June 20, Kathy wrote in her journal that at 7:00 a.m., she rotated the eggs 180 degrees and egg number 6 talked back! Sure enough, when I held the egg to my ear, I heard a faint "squeak, squeak." With this hopeful sign, we decided to stay with the bacteria incubator.

Each tiny achievement with the guillemots brought me closer to believing that I was learning information that would somehow help with puffins. After the first "squeak" came vibrations from within the blue shell, and a closer inspection of the dark chocolate splotches revealed a tiny star-shaped crack. We held the egg to our ears like expectant parents and greeted every squeak with big smiles. Maintaining the temperature at 93–98 degrees seemed ample at this stage of incubation to keep the egg on track for hatching.

A few hours later, miraculously, the tiny guillemot's egg tooth (a hard calcium ridge on the tip of the upper mandible) emerged from the starred pattern and the chiseling (and squeaking) continued. Before we knew it, the first wet guillemot tumbled from its shell. Kathy put on clean cotton gloves and took the soaking wet chick out of the incubator to transfer it to a cardboard brooding box fitted with an infrared lamp to warm and dry our small charges, and a cover for darkness. As midnight approached, she went back to the incubator to discover a second chick with eyes wide open and half dry. We transferred it to the brooder box. We were thrilled to see that both chicks were active and strong.

The next morning, as two more eggs started hatching, we started feeding the first two chicks slices of thawed hake with sterilized tweezers. By afternoon both birds were waddling around. On the advice of Joe Bell, from the Bronx Zoo, we started supplementing their meals with vitamin E, B-1, and multivitamin pills, which we pushed directly into their mouths or hid in the fish.

By June 23, we were up to six chicks. At first some were more awkward than others, falling helplessly on their backs. Each time, we gently picked them up and placed them back on their feet. With time, they became increasingly active and ate well, and we started to think we knew what we were doing.

June 25 brought our first sadness. Our second oldest chick died at four days, probably of overheating from being too close to the heat lamp. We installed a heat shield made from blue corduroy cloth. By June 29, three more eggs had hatched, bringing our fuzzy little flock to eight.

Then, on July 1, we moved the incubator with the remaining eggs from the common interior room of the Binnacle to the porch, which I had insulated and equipped with an air conditioner. My fellow Hog Island instructors were glad to be relieved of the stink from my growing zoo, although they never complained because they were bird enthusiasts too and were fascinated with our evolving process.

This porch would eventually be the puffin chick-rearing room—my humble parallel to the "peregrine palace" at Cornell, where Tom Cade kept peregrines until they were ready to be placed in hacking towers for eventual release. But the Binnacle was no palace by any stretch. It was a modest, unheated three-room cabin, with an add-on porch that I had now claimed to rear the chicks.

Despite the loss of one guillemot chick, all was going well until July 19, when disaster struck. An electrical storm swept over Hog Island, knocking out the power at 3:00 a.m. Because a constant warm temperature was critical for the project, we scrambled to gather all the chicks, the brooder box, the remaining eggs, and the incubator, and boated back to the mainland, where there was still electricity. We were able to return to Hog Island by dinnertime, but the sudden chilling killed the remaining eggs. The next day, Kathy noticed the chicks had lost their appetite.

The storm, loss of the remaining eggs, and uncertainty about the condition of the surviving guillemot chicks underscored the uncertainties that lay ahead, but somehow we managed to stay optimistic and adjust our methods as needed.

We knew from the outset that we would need to release the puffin chicks at one of the Egg Rocks, and fortunately we had a boat that we could use to explore the islands and determine which one would be most appropriate. I named the seventeen-foot Aquasport outboard the *Lunda,* which was then the genus name of the tufted puffin. The Egg Rock shorelines are so rugged that a small landing boat was necessary to transfer from the Aquasport to the island, so I purchased a "sportyak," a lightweight but rugged plastic rowboat that we named the *Geezer*—a nickname that we had come up with for a low-floating guillemot.

With boat and landing skiff at my disposal, I proposed an overnight campout on Western Egg Rock in July. I invited Kathy Blanchard, Hog Island botanist Grace Bommarito, and David Morton, Dur and Peggy's teenage son. It was late afternoon before we had the boat stocked with all the necessary camping equipment, but we were excited about the adventure as we groped our way through the late-day fog that hid most of the landmarks on our way out to the mouth of Muscongus Bay. By the time we were in sight of Western Egg Rock, we realized that the motor could barely push us forward, and we were riding surprisingly low in the water.

We made it to the north end of Western Egg Rock but snagged a lobster buoy that turned the *Lunda*'s stern into the incoming swells. Before we knew it, water was pouring over the stern and leaking into the hull, which was already soaked from the trip out. Within minutes, the boat started listing and water began to cascade over the gunnels. With everyone bailing madly, I put the *Geezer* over the side, hopped in, and pulled myself around to the stern, hoping to untangle the lobster trap. But there was far too much tension to untangle the rope from the propeller. There was no time to waste; I had to cut the line.

Once we were free of the trap, I climbed back onboard, taking care to keep our weight centered while we frantically bailed. We were relieved to get the ancient motor going so that we could position her for anchoring. I rowed the crew ashore one at a time, but each trip added more water to the tiny skiff, which felt more like rowing a bathtub than a landing craft. Eventually, I made the last shuttle from the *Lunda* to shore, where David and Kathy grabbed the skiff and together we pulled it up onto the island over the slippery boulders that felt like so many greased bowling balls. Later, Grace mentioned that she didn't know how to swim and so was especially happy that I got her to shore.

We managed to start a driftwood fire. Everything was soaked, including sleeping bags and tents. But the novelty of being perhaps the first people ever to camp on Western Egg Rock diffused our discomfort. After a soggy dinner of oatmeal, we pulled out flashlights that somehow still worked and walked around. The big surprise was how many great blackbacked gulls were on the island. They behaved as if they had

David Morton (left), Steve Kress, and Kathy Blanchard on Western Egg Rock in July 1973. We were soaked from a harrowing landing but exhilarated by our first adventure of camping on a bird island. Photo by Stephen W. Kress

never seen humans before. Normally the gulls roost on the water at night, but perhaps because of the fog, these were holding their ground, permitting us to approach within a few feet as they sat blinking in the foggy darkness. The gulls that did fly off began a deep, resonant, haunting chant that only a great black-backed gull can understand: "HA HA HA!"

We soon realized that Western Egg was probably a poor home for a puffin restoration project. The boulders were smaller than those I remembered at Machias Seal Island, and I suspected they would not support many puffins. I also worried about the treacherous landing, knowing that people would need to come and go from the restoration site with some degree of safety.

Suddenly, it was only two days before our scheduled return to Great Island, Newfoundland, to collect the six puffin chicks. Although we had several surviving guillemot chicks, their ups and downs hardly represented a gold standard of nurturing. Kathy frequently had to wipe off their feet and beaks, which they constantly soiled on our newspaper "carpets." We wondered if anything we were learning was worthwhile for the high-stakes challenge to come.

We avoided that question by staying busy with plans for the six puffin chicks. We decided to rear three of the puffins in the air-conditioned Binnacle porch in wooden boxes with hardware cloth bottoms through which we hoped the guano would drain sufficiently to prevent soiling the chicks. The other three were to be reared in rock burrows, built on the shore of Hog Island with scavenged chunks of granite and cement. The rock burrows more closely replicated the conditions of natural puffin burrows, and I was already thinking that replicating natural burrows would be better than rearing chicks in artificial cages.

I also hoped to feed the chicks freshly captured fish of the right size but soon realized that it was easier to build a burrow than to capture fish. My first approach to obtaining fish was to set up a large fish trap modeled after a herring weir like the ones used in the Bay of Fundy. Nothing was too big or ambitious to undertake if it would help us successfully rear the chicks. Kathy and I designed on paper a fish trap with two net-covered wings that would extend across Hog Island's hundred-foot-wide Long Cove. My ever-hopeful vision was that the trap's wings would funnel small fish into a circular, mesh-walled trap at high tide; at low tide we could then go out and reap our harvest by using a dip net to scoop fish from the trap.

With the help of Hog Island campers, we cut down about

two dozen spruce saplings, trimmed off extraneous branches, and pounded the posts into the mud at low tide. We then hung fine-meshed netting to create the wings and circular trap. I envisioned a glut of perfectly sized fish, ideal for feeding the pufflings. But like most of our endeavors that summer, the Hog Island weir was long on energy and hope but short on research and experience. Much to our disappointment, our contraption captured just a few tiny fish before the netting became draped with heavy seaweed, collapsing most of the structure.

Abandoning the weir idea, we cut the netting down to create a large seine that we could pull through the water, similar to the way I had collected minnows and darters with Mac Albin as a kid in Ohio. That method yielded better results, but live fish were not a dependable source, and we soon realized that if we were ever to bring the project to scale, we would need to find a reliable supply of frozen fish and deal with the vitamin deficiencies.

Kathy scouted the neighborhood waterfront during the days just before our trip to Newfoundland, asking local fishermen where we could find a dependable supply of small fish. We could not find herring smaller than about six inches, which was much too large to feed puffin chicks. Eventually, her inquiries led her to a fish locker in nearby Waldoboro that could purchase and store blocks of frozen smelt.

While Kathy was searching for fish, I put the finishing touches on my carrying case for transporting the chicks. The frame was a fourteen-by-eight-by-ten-inch plywood box with a burlap back wall and swinging plywood/burlap doors on the front for access. Inside were six juice cans. We spread silicone caulk on the bottom of each can, and sprinkled sand into the tacky surface to provide a good grip for the chicks' feet. Kathy deemed it in our journal notes to be a very practical and safe carrying case.

I was worried about making another long, arduous trip with the chicks over rugged Newfoundland roads and shared this concern with Peggy Morton. Always resourceful, Peggy began thinking about options, and just days before our planned departure, she introduced me to Bob Noyce, a mainland neighbor with a sense of adventure and a man who knew what it meant to develop new ideas. Indeed, this was Robert Norton Noyce, cofounder of Intel Corporation. Along with Jack Kilby, he invented the integrated circuit microchip that fueled the personal computer revolution and, with Gordon Moore, was a founder of Intel. Bob Noyce eventually held fifteen patents for computer microchips and was honored by three presidents, but in 1973 I found him modest about his interests and happy to help with my adventure. He offered to fly me and Kathy to Newfoundland in his private jet along with a couple of neighbors whom he rounded up as helpers.

The flight eastward along the Maine coast and then over the Maine and New Brunswick forests was stunning—so many islands, so many lakes. Until now, I had no idea what a migrating bird might see over this magnificent coast.

We stayed overnight in a hotel in Saint John's, and early the next morning we all headed down to East Bauline to meet John Reddick. Bob Noyce joined us for the trip out to Great Island, and for the second time that amazing summer we grabbed the swaying ropes over the Great Island chasm and pulled ourselves up onto its guano-decorated rocks. Collecting six chicks would take only a few hours, so John waited offshore and then returned us to the mainland with our prize.

I had not only a suitcase of puffins but also "starter food" in the form of two ten-pound bags of capelin that I shoveled off the beach. To the casual onlooker, capelin is just another small silver fish, but it is the critical forage fish—the vital bridge between zooplankton, marine mammals such as whales and seals, commercially valuable fish such as cod, and most Newfoundland seabirds, including puffins.

I kept the chicks by my side on the plane while Bob whisked us back to Maine on Sunday, July 15. At last, Project Puffin was really beginning.

When we arrived back at the Audubon camp, I carried the six chicks on the final leg of their journey to the Binnacle. We transferred the chicks to boxes and fed them some of the capelin. Most ate well, but we were concerned that the lightest one showed little interest in food. We took its temperature by inserting a rectal thermometer into its cloaca (the opening through which birds eliminate waste) and found it was 99.2°F, but we weren't sure what was normal. We reasoned that even if we were not sure what normal temperature was for a tiny puffin chick, a change (preferably a little higher) would be a good sign.

The next morning we moved the chicks to their new homes. We placed three chicks in separate rock burrows and covered the entrances with wire mesh. These were located above the high-tide line on the shore below the Binnacle. The remaining three were placed in plywood chick-rearing boxes on the porch. Although it was my intent to evaluate the two rearing methods, I favored the outdoor rock crevices not only because they were much more like a natural burrow but also because I couldn't imagine building a hundred or more of the boxes and the time it would take to clean them, to say nothing about the size and cost of an air-conditioned facility to house a hundred or more chicks!

The Hog Island campers were beside themselves with curios
ity, so I brought one of the fluffy black chicks in its box outside
so that some fifty adult campers could see it. The chick huddled in the corner, emitting a whiny, sirenlike call as a long line
of campers cued up and waited their turn to peek in the box,
each whispering good wishes to the fluffy chick like so many
wise men who had come to see the baby Jesus.

But our excitement was short-lived. The day after the
transplant, one of our three puffins from the rock burrows was
missing! At first we suspected our own shoddy handiwork. We
discovered a crack in the burrow large enough for the chick
to escape. We immediately applied cement to the crack and
checked every nearby crevice to find the breakout chick. Alas,
we found nothing.

Sober with regret over losing one of our precious puffin
chicks, we decided to entomb the remaining two chicks in
their burrows by cementing hardware cloth over the burrow
entrances, leaving just a tiny opening through which we could
slip in the thawed smelt. We replaced the lost puffin with a
home-hatched guillemot hoping to learn if a rock crevice habitat would give it better waterproofing.

Then more panic.

On Friday, the guillemot chick also vanished from the
same rock burrow. We were desperate for an answer to this
mystery. We suspected rocks caving in due to a careless hiker
or more shoddy craftsmanship on our part, but then we noticed a telltale clue—a few raccoon hairs on the hardware cloth
doors. We brought the remaining puffin chicks back indoors.

As August approached, fledging time grew near for the
guillemots. Remembering the trials of the previous summer,
we thought it best to test the waterproofing on our chicks. We
gave them a test run in the tub, one chick at a time. It took just

a few minutes to see that the first chick became waterlogged and sat low in the water. We tested the other guillemot chicks and found the same disappointing results.

We noticed, however, that the chicks began preening when they got wet, so we hoped that this would help them clean and rearrange their feathers. We modified the tub by building a ramp at one end with a heat lamp over a shelf so that the chicks could climb out of the water and dry off. But this and other efforts to help the guillemots develop waterproofing failed.

On July 30, we decided to test the rock burrows again with another guillemot and two of the remaining puffin chicks. Knowing the risk from raccoons, we doubled our security efforts by cementing hardware cloth over the burrow entrances and built a chicken-wire cage over the burrow complex. We also set a Havahart trap on the rocks for the suspected raccoon but caught nothing.

To help the surviving guillemots acquire waterproofing, we constructed a large chicken-wire pen at the edge of Porthole Cove adjacent to our rock burrows. Twice-daily tidal changes would give the birds plenty of ocean water to help them clean feathers and stimulate preening. The pen was about fifteen feet long and six feet wide and stretched from the high-tide mark to not quite the low-tide mark. We included a granite outcrop to provide ample resting places where the birds could preen and dry off. We placed the three remaining guillemots in the pen. Soon they were swimming and preening with improved waterproofing. Forever optimistic, we hoped that we had resolved the waterproofing issue.

The guillemots adapted well to their new home. There were even signs of improved feather condition. Kathy camped out by the pen just in case the raccoon made a brazen raid, but

just four days after placing the chicks in the pen, she awoke
to discover that the raccoon had breached our security and
somehow killed all the chicks. Kathy found me in the Hog Island office. In tears, she held the pathetic remains—a couple of
guillemot legs—that still held the bird bands we had affixed. It
was the saddest day of the summer.

Guillemots seem like hardy birds, but if we had been attempting to restore guillemots rather than puffins, this waterproofing problem would have brought the program to a halt before
it started. It was indeed sobering to compare our captive-reared
guillemot chicks with wild chicks that were perfectly clean and
exhibiting wild, feisty vigor.

I wondered if our handling was the cause of the waterproof issue. Perhaps oil from our hands was interfering with
the natural waterproofing of the feathers. Later, I learned that
aviculturists at aquaria with guillemots and other alcids avoid
handling seabirds with bare hands and usually use a net or
gloves when it is necessary to move or examine birds. Or was
the problem nutritional, since we were feeding thawed fish
strips that oozed fish oils onto beak and breast rather than the
guillemots' usual diet of whole rock eels? Or was it exposure to
the chicks' oily excrement? Or was it the insufficient sunlight
that prevented ample exposure to vitamin D and the development of the uropygial gland that produces normal preening
lubricant?

Like most of the problems we encountered, all we could
do was keep trying our best guesses. Yet the stakes were high,
because I knew that the chances of getting more puffins in the
future would be dismal if we failed with our first chicks. I concluded that, if puffins were as difficult to rear as guillemots, we
were really in trouble.

With fledging time for the Hog Island puffin chicks approaching, I mounted another scouting trip—this time to Eastern Egg Rock. I invited the camp's intrepid botanist, Grace Bommarito, who was brave to consider another boating adventure with me. My goal was to see if Eastern Egg Rock would be a better location for the project than Western Egg Rock. We took the *Lunda* out, dropped anchor off the north end, and paddled ashore with camping equipment for an overnight stay.

I was pleased to find habitat that looked much more like Machias Seal Island, with larger boulders piled in great stacks around its perimeter. A few of the boulders were of glacial origin, dropped thousands of years ago when ice fields retreated from Maine, but most were more recent, snapped from the underlying bedrock and then rolled into place during extreme storms. The net result was a natural apartment complex with many layers of burrows under the jumbled boulders that was well above the high-tide mark. Grace and I discovered a few Leach's storm-petrels nesting under driftwood and several dried and twisted skins of black guillemots turned inside out. This was telltale evidence of attacks by great black-backed gulls. Despite these casualties, it was good news that guillemots were nesting here. I reasoned that if they could still breed here, then so could the puffins. I could imagine the puffins standing against the blue sea, flying in with loads of glistening fish, dropping into dark crevices to feed chicks.

The dream was all so clear, and Eastern Egg Rock seemed so perfect, that it did not matter to me that the next morning, when we attempted to turn over the decrepit outboard, it made an awful clunk. Later, I learned that we had heard the death sound of the motor—a rod was thrown and the motor was ruined. Eventually, a kind lobsterman towed us back to Hog Island.

I was puzzled why puffins had not naturally recolonized Eastern Egg in the nearly nine decades since the last documentation of their presence. Puffins were a protected species under the Migratory Bird Treaty Act of 1918, and Eastern Egg saw few human visitors to disturb the recolonization process. Although most young puffins return to their natal home, puffin colonies also naturally start on their own. Why had puffins not naturally recolonized Egg Rock?

On August 16, we began to notice that the puffins in the Binnacle were becoming restless, fidgeting frantically. Clearly it was fledging time.

The raccoon raids that decimated the guillemots and threatened the puffin chicks were a clear message that I could not rear large numbers of puffins on Hog Island and later release the fledglings at the puffin-nesting site. We needed to move the entire chick-rearing and fledging process to Eastern Egg Rock.

In the days before we moved the puffin chicks, Kathy and I had hauled bags of cement and collected rocks to build five release burrows on the southwest shore of Eastern Egg Rock. On August 17, I placed the remaining chicks in a box with separate compartments for each puffling. The box had a burlap-covered top for plenty of ventilation. Wick Campbell, second boatman for the Audubon camp, took us out in the camp's thirty-five-passenger *Osprey III*. Joining Kathy and me for the momentous trip were Peggy Morton, her teenage daughter, Leslie, and Karen Thom, the wife of Hog Island instructor Rick Thom.

We arrived in late afternoon, and Kathy, Karen, and Leslie promptly set about putting up their tents while Wick began setting up a portable CB unit that was to be their primary means of communicating with me back at Hog Island. Because

Hatched in Newfoundland and reared in experimental burrows on Hog Island, the first five puffin chicks were placed in a transport box for their boat ride to Eastern Egg Rock for release on August 17, 1973. Photo by Stephen W. Kress

I was still teaching full-time in addition to all that I was doing with puffins and guillemots, I would not be able to join them on this camping trip.

I banded each chick with a green-coiled plastic band on one leg and a metal USFWS band on the other. Each was placed in its own rock crevice burrow, and wire doors were secured over the entrances to give the chicks a chance to settle in until dark, when the covers would be removed. Wary from the raccoon raids at Hog Island, we now feared that gulls might swoop in and carry off the chicks, so we waited until midnight, when most of the gulls would be asleep, to open the burrow entrances. With the tent camp set up and the pufflings settled into their temporary homes, I returned with Wick and Peggy to Hog Island, where we would wait for news.

That night, Kathy, Leslie, and Karen took turns watching the rock burrows, ready to chase off marauding gulls and guarding against high tides. With the burrows open, the chicks were free to explore outside and make their way to the sea when ready.

To view the chicks' progress, the women cloaked their flashlights with red cellophane. In this low-tech operation, that was our best approximation of night-vision equipment. Over the next three nights, Kathy, Leslie, and Karen stayed up to watch the shadowy shapes come out of the burrows, exercise their wings, and then flap and flutter their way over the rocks, drawn by an instinctive attraction to the sound of the surf and contrast of light on the horizon.

Kathy's journal notes document the excitement of seeing the fledglings make their way to the water. Most of the chicks walked, climbed over boulders, and eventually scrambled into the surf.

But one of the pufflings made a more dramatic departure. On August 19, Kathy described in her journal how she watched puffin number 3 leave its burrow on its way toward the sea. The hardy chick moved slowly, in a straight path toward the water avoiding most of the cracks and ravines. Sometimes, it would tumble out of sight, then scramble back to the surface. Finally, it clambered up a tall rock overlooking the sea, where it stood for about five minutes (as if trying to get the nerve to jump). Several times it leaned forward as if to jump, followed by rapid beating of wings. Then, twenty-five minutes after leaving its burrow, IT FLEW into the air above the crest of the waves and disappeared into the darkness. Kathy wrote that she was "thrilled and honored to witness this special moment!" Considering all that puffling had been through, seeing it disappear on its own winged flight was indeed special.

We had banded the chicks with the hope that we would someday see them again, but I knew these chicks were likely a sacrifice for science because only a fraction of fledglings survive the rigors of the sea. Our primary purpose in this first year was to simply prove that we could translocate chicks from Newfoundland to Maine and rear them to fledging age. The question was whether we proved it enough to earn the chance to bring many more chicks in subsequent years. The puffin chicks looked cleaner than our guillemots, but we could only hope that they had better waterproofing. There was no way to know for sure.

At summer's end, I reported to Nettleship the successful fledging of the chicks and my greater appreciation for the stamina of puffins. Fearful that the sometimes-chaotic nature of our seat-of-the-pants approach might leave a negative impression, I neglected to mention the trials with guillemots, raccoons, waterproofing, the collapsed weir for capturing fish, and near disasters on boats. Fledging the chicks from Eastern Egg Rock was the high point of the summer, and it set our path and vision of the future. I needed Nettleship's endorsement for an expanded program the following summer. Remembering Bill Drury's wisdom, I asked for his advice on future steps. About two weeks later, he wrote back, giving his support for more chicks the next summer. I could have walked on air.

These first five birds were our victory—they represented an enormous amount of effort, trial and error, and no small amount of dumb luck. The five puffins were the first of nearly

two thousand that would follow a similar path from Newfoundland to Maine. Each was a miracle and offered hope that
someday puffins would return to Egg Rock to found a new
colony.

5

Massacre to Miracle

We approached Eastern Egg Rock on June 14, 1974, ready to take on whatever the largely unknown process of restoration dictated. It was a beautiful day: fair and mild, with calm seas. Joe Johansen, the camp's new head boatman, was at the helm of the *Puffin III*, which was loaded with an odd assortment of gear for rearing puffins, lumber for a tent platform, and camping equipment. Kathy was returning for a second summer as my assistant, joined by new research assistants Nina Davis and Roger Eberhardt. Dur Morton, who was now director of the camp, joined us for the trip. Also onboard was Audubon chronicler Frank Graham Jr. and photographer Chris Ayres, on assignment from *Audubon Magazine*.

Onboard with us was one other person, Frank Gramlich, Maine's federal supervisor of wildlife services. Before we could bring puffin chicks to the island, we had to displace the long-established colony of herring and great black-backed gulls. I had come to realize that these large gulls were the primary reason that puffins had not naturally reclaimed Eastern Egg Rock in the decades since their demise. Puffins instinctively

recognize gulls as fierce predators and shy away from colonizing otherwise good habitat when they see a legion of threatening gulls. The US Fish and Wildlife Service told me that the
gulls would have to be poisoned, and there was no one better
for the job than Gramlich. At the time, the primary approach
to gull management was the use of the avicide DRC-1339.

Gramlich was a legendary and complicated figure in the
tiny world of wild animal control. He was a World War II
Battle of the Bulge and Korean War veteran who had loved
hunting rabbits and ducks while growing up on eastern Long
Island. When he retired from the military in 1959, his fascination with animals led him to biology as a second career. With
just one year of high school education, he successfully took a
placement test and entered the University of Maine, where he
earned a master's degree in wildlife management.

After graduating, he waded into the thorny side of his
craft: when and where to kill some wildlife to preserve others.
He also found himself wrestling with when it was appropriate
to kill wildlife for human comfort and commerce. In the late
1960s he field-tested chemical control of urban pigeons in
Bangor; in the 1970s he spearheaded federal coyote control
efforts to benefit farmers in the West.

But he was also a preservationist. The year before Gramlich joined us on the trip to Eastern Egg Rock, a Maine state
representative had proposed bounties for coyotes, which farmers were blaming for livestock deaths. As a *Maine Times* story
noted, the bill was killed by "an elderly, weather-beaten gentle-
man named Frank Gramlich, Maine State Supervisor for the
Division of Wildlife Services—the very government agency
which had for so long been entrusted with the attempted offi-
cial extermination of the coyote in the West."[1]

Even as he boated out with us, Gramlich's mind was on

bald eagle nests. Maine's eagles were too saturated with DDT
to consistently produce eggs with shells thick enough to withstand incubation. Earlier that spring, Gramlich's fellow researchers had taken some failed Maine eggs out of their nests
for testing. One of them turned out to have the highest known
concentration of DDT from an eagle egg anywhere in the United
States. As a result, Gramlich had launched a program of bringing eggs from much healthier eagle populations in Minnesota
and climbing up into the trees to substitute them for the likely
defective Maine eggs.

So in a very real sense, he was a fellow traveler in restoration. When he was once asked why he chose wildlife management as his postmilitary career, Gramlich said it was wrong to
destroy creatures to the point where they cannot be retrieved.

He came with us to Eastern Egg to destroy a portion of
the bird population so that another part could be "retrieved."
He was armed with a brown plastic bag containing six hundred pieces of bread smeared with Starlicide-treated marga-
rine. The poison's name derived from its first target, the starlings and blackbirds that swarm on chicken and livestock feed.
Chemists had discovered that gulls were also highly sensitive
to DRC-1339. The poison shuts down the birds' kidneys, leading to death within three days. Starlicide was recommended
because it does not hydrolyze into dangerous compounds in
water and is not thought to threaten scavengers such as eagles
that might eat a sick gull.

This was actually Gramlich's second trip out here this
year. His first visit bringing Starlicide sandwiches was back in
May. He estimated at that time there were about a hundred
gulls to deal with. He put two or three death sandwiches on
each nest. "The gulls were taking them just fine," Gramlich said.
"I figured that would clean them out."

When we arrived on Egg Rock at 8:11 that morning, Gramlich exclaimed, "Look at them! There must be three hundred, maybe four hundred around the island now!"

Black-backed and herring gulls rose in great clouds as we landed, chanting their base alarm call, "HA HA HA," which I was coming to know very well from my many intrusions into their colonies. There were active nests at our feet, and the parents were intent on chasing us off.

This was my first trip with Joe Johansen, a modest and muscular man who was a seasoned salt with an endless nautical vernacular. He had recently retired from a career with the US Coast Guard maintaining equipment in remote lighthouses. His arrival was great good fortune for me, Project Puffin, and the Audubon Camp. Joe, a master oarsman, would stand upright, rowing the twenty-one-foot wooden dories with great skill. He and his dear wife, Mary, became great friends of mine, and more than a few times Joe rescued me from boating mishaps. When the time came to land puffin chicks on Egg Rock, I could count on Joe to land the chicks safely ashore (usually at night) for the next two decades.

Johansen rowed us ashore with his "Norwegian steam," and we began the endless process of "lumping," as he called hauling by hand. As we sorted out the materials for burrows and bird blinds, Graham and Ayres followed Gramlich around the island on his grim task. With his toe he poked a few of the long-dead gulls from his first visit in May. He smashed new gull eggs by stepping on them. And then he laid the Starlicide-treated sandwiches in the nest. He explained to Graham, who was taking notes for a book he was planning on humans' chaotic relationship with gulls, "If the gull eggs are broken, they'll come back and eat them. And they'll gobble the bread right along with the eggs."

Gramlich admitted to Graham, "This control program is no solution to the gull problem. It's only a stopgap. But something had to be done right away to save the other species from the terrific predation on their eggs and young. We included this island on an experimental basis because of the puffin project. Puffins couldn't get a foothold here with all these gulls. Look at this!" Gramlich and Graham peered down at the red feet and feathers that were all that was left of a black guillemot, probably devoured by a black-back.

Gramlich crunched the eggs on more than two hundred nests containing more than four hundred eggs, some of which were near hatching. "A hell of a way to make a living," Gramlich said. And there was collateral damage from our presence. As Gramlich worked his way around the island, he bypassed sixty-five double-crested cormorant nests, most of which had eggs. But the disturbance flushed the cormorants off, permitting gulls to descend and eat the eggs.

Kathy, Nina, Roger, and I returned to the island the next day in the *Lunda,* tied up to the mooring that Joe had dropped yesterday and rowed ashore in the *Geezer*—to set up camp. Nina was a natural for the job, with a mind for detail and a nurturing persona. Years later, she traded her field biology career for a profession in nursing, which somehow seemed appropriate considering the TLC that I saw her bestow on puffin chicks. Roger was the younger brother of Ellen Eberhardt, marine life instructor on Hog Island during these years. He was an undergraduate at Earlham College. Soft-spoken, with a ready smile, he was handy with rope and tools and willing to do most anything for wildlife, a natural for island life.

Soon we would return to Great Island to collect fifty puffin chicks. This time, instead of raising the chicks at Hog Island, we planned to boat them straight out to Eastern Egg

Landing at Egg Rock in 1974 was accomplished using the *Geezer,* a plastic skiff designed for calm waters. Here the 1974 Project Puffin team unloads supplies at low tide. Left to right, Kathy Blanchard, Roger Eberhardt, Steve Kress, and Nina Davis. Photo by Duryea Morton

Rock. We would feed the chicks freshly thawed smelt and vitamin supplements three times daily until they reached fledging age. We would close the burrow entrances by day to prevent predation by gulls and open them at night to permit the growing birds to exercise outside their burrow entrances. My hope was that this would help the chicks learn that Egg Rock was home and reduce the chance that they would return to Newfoundland. We would also band the chicks with metal Fish and Wildlife Service bands and colored plastic leg bands so that we could recognize them on their return.

First order of the day was digging a latrine, followed by assembling the burrows and building our tent platform. Like homesteaders, we brought a few plants from the mainland and dug a little garden in the peat for lettuce and beans. Next we needed to cut a trail from the landing to the campsite, about two-thirds of the way across the island. Although it would have been simpler to establish camp near the landing, I wanted to be near the puffin burrows, which we would build on the southeast end of the island, where the chicks could readily find their way to the water. So the lumping began; trip by trip, we slowly hauled up the mountain of wood for the twelve-by-twelve-foot tent platform, materials for burrows, water jugs, and camping gear. The work was balanced by the sense of adventure and anticipation for the arrival of puffin chicks.

We took a break for a bird count. Great black-backs, not yet affected by the poison, still dominated the island. As we scanned, we saw that the rock burrows for our five puffin pioneers were gone, likely washed away by winter storms that blasted seawater across the island and lifted granite boulders weighing many tons. Before retiring that night, we destroyed many more gull eggs and some chicks. Without Gramlich, this grim task had passed to us.

We started making the "burrows" the next day. There were few small, manageably sized rocks on Eastern Egg, so I had settled on a design that used ceramic chimney liner tiles to replicate stone burrows. Each tile was about eight inches square and eighteen inches long. We strapped three tiles together with wooden frames to create homes for three chicks—one chick for each tile home. We then tucked the artificial burrows among the natural granite burrows, positioning them on flat rocks from which we envisioned the puffins fledging easily into the sea. The backs of the burrows were finished with wood, and we fashioned hardware cloth doors over the front. We reasoned that one practical benefit of the tile burrow sets was that we could move and stack them away from the shore in winter.

After three days we completed twenty-two burrow octs cach with three ceramic tile burrows. These would accommodate the chicks for the restoration project as well as ten chicks that were part of a special project with the Mount Desert Island Biological Laboratory.

As we built the burrows, we kept a wary eye on the gulls. On the fifth day after Gramlich served his Starlicide sandwiches, we started to find dead birds—twenty-five black-backs and four herring gulls.

Now a week into construction, we mixed concrete to build ramps in front of the burrows so the chicks could make their way more easily to the surrounding granite. We wanted to do everything possible to make the burrows suitable but later realized how ridiculous it was to create little ramps for the pufflings, considering that they would soon be scrambling over huge boulders on their initial journey toward the surf. We were so caught up in the details of preparing for the precious puffins and so proud that we had invented what we thought were the perfect puffin homes that it was hard to see what was coming.

Three weeks later, on July 11, Kathy and I met up with Gramlich and his assistant Gary Donovan. We boated out to Eastern Egg Rock. This time, instead of going on land, Gramlich tossed Starlicide sandwiches into the water, but few of the gulls picked up the bait. Gramlich suspected that the gulls recognized him as the agent of death and so avoided anything that he delivered. He said he wanted to try something else on his next visit.

It was so clear on July 12 that we could see eight lighthouses on the horizon, including, we thought, Matinicus Rock, thirty-five miles to the east, where a single pair of puffins—the last survivors of Maine's original population—nested in 1902.

The trail we had cut to our camp was already overgrown and had to be cleared again with a scythe. But we were happy to see that our garden was also growing. The lettuce and beans had grown three times their former size, no doubt well fertilized by seabird guano.

On this perfect day with southwesterly breezes, we started putting the hardware cloth over the doors of the puffin burrows and naively thought this would be the last detail to create perfect puffin homes. We also set up our weather gear to measure wind, air temperature and pressure, and rainfall. All was going well until we started seeing temperatures in the tile-and-wood burrows that concerned us. In the heat of the afternoon, the temperature in our ceramic burrows was 80°F. The Newfoundland puffins would be coming from cool soil burrows with a temperature in the 50s. With just a few days before the transplant, we cut thick chunks of sod and placed them on top of the ceramic burrows to provide insulation from the summer sun while Nina prepared a hand-dug earthen burrow as a possible alternative.

The next morning we hoisted our puffin flag high over a knoll on the south end of the island, a location we still call Flagpole Hill. With the Newfoundland trip now imminent, Joe Johansen brought out Leslie Morton to stay with Nina while Kathy, Roger, and I went back to Hog Island to prepare for the trip to Great Island. The relief crew continued painting the ceramic burrows, making the interior black to simulate the dark interior of a soil burrow; they painted the outside light gray in the hope that it would have a cooling effect. But temperature problems persisted; the tile burrows were now registering 92 degrees at noon. By contrast, Nina's hand-dug soil burrow measured just 68 degrees.

On July 15, the day I was to leave for Great Island, Nina and Leslie crushed vitamin pills and placed the bits into sterilized bottles in preparation for feeding the chicks. And they kept trying to dig earthen burrows, but it was difficult work because of buried rock. With granite bedrock just inches below the hand-dug burrows, we worried about the threat of flooding following heavy rains.

To move the puffin chicks over the Canadian border, I would need permits from the US Fish and Wildlife Service and the Department of Agriculture (USDA), as well as export permits from the Canadian Wildlife Service and the Province of Newfoundland. To move and band the chicks, I eventually obtained thirteen permits for the project—ironic because wild puffins fly over the same border without a scrap of paper. Of course the permits were the result of the puffin's protected status, which they received along with most other birds as a result of the early twentieth-century bird protection movement.

The most challenging permit was the one from the USDA. This agency was involved because of a recent outbreak of Newcastle disease, a deadly poultry virus. My contact for this permit was Dennis Decoteau, a USDA veterinarian based in Massachusetts. At first his office insisted that we place all the birds in a USDA quarantine facility, but this was not acceptable. Quarantine might work for chickens and ducks, but not puffins. I was concerned that the chicks would suffer waterproofing issues from stressed living conditions, and the details of feeding were largely unknown. We eventually agreed on a solution: Eastern Egg Rock would serve as the quarantine facility for most of the chicks, and four chicks would be held at Roger Williams Park Zoo in Rhode Island for observation in case they were infected. With a solution for the inspection and with permits in hand, I was looking forward to returning to Great Island. Our pilot was again the generous Bob Noyce.

But just before we were to leave from Maine to collect the chicks, Bob called to inform me about a complication. He explained that his wife, Betty, wanted to come on the trip with us—but she was allergic to birds! He would fly me to Newfoundland as agreed, but Betty's allergy meant that we would need to place the chicks in the baggage compartment of his plane rather than have the birds in the seating area with us. Alas, his baggage compartment was neither pressurized nor heated.

Bob said that he could fly at ten thousand feet, where there would be little need for regulating air pressure, but he was not sure what the temperature would be in the baggage compartment. He hoped to test this before the flight with a hardware store–quality high-low thermometer to give us a sense of what to expect, but he never found the opportunity to run the test.

I was very concerned about what would happen if we were forced to sit on the runway awaiting a customs agent. If we were delayed on the ground on a sunny day, the baggage area would heat up, and the chicks would likely suffer or perish. I didn't have time or the resources to hire another pilot; we needed to move forward and take our chances.

We met Bob at the Wiscasset airport, a half-hour drive from Bremen. We loaded three puffin cases, each with twenty cans, and a smaller case to hold the chicks intended for the USDA. We circled Eastern Egg Rock and then flew up the coast to Newfoundland, about a thousand miles to the north and east.

Our flight to Saint John's was uneventful, but I worried about the temperature in the baggage compartment. On arrival, we checked the thermometer and found that it registered moderate temperatures, which gave me some relief—but we had spent little time sitting on a runway in full sunlight.

In Saint John's, we checked into our hotel, and Kathy and I walked over to the office of Leslie Tuck, where we were to pick up our collecting permits. Tuck was more staggering proof how few degrees of separation there were in the bird world. He'd attended Harvard University from 1936 to 1938, where he was a student of the same Ludlow Griscom who mentored Irv Kassoy and the Bronx County Bird Club.

Bob and Betty opted to stay in Saint John's while Kathy, Roger, and I woke up well before dawn on July 17 and drove our rental car about an hour south to the sleepy village of East Bauline, where we met John Reddick for our return to Great Island. The seas were rougher than a year earlier, which meant rowing ashore in a small dory rather than John's "one-lung" powerboat. By 6:30 a.m., we were scaling the cliffs of Great Island once more.

In his characteristically thorough way, David Nettleship had provided a map with specific areas for the collecting. This year we were to take chicks between thirteen and eighteen days old, with a specific range of wing lengths of 1.5 to 2.5 inches as measured on a ruler from the bend of the folded wing to the tip of the longest primary feather. We believed that chicks of this age could maintain their own body temperature and had the greatest chance of learning a new home when translocated to Maine.

A cornered puffin is not something to take lightly, despite its cuddly appearance. Its beak has a sharp, almost raptorlike tip, and a puffin is not shy about planting that crusher beak around a finger or sinking the tip into the loose skin between a puffin grubber's fingers and hanging on with great determination. To collect our quota, we probed more than six hundred burrows. About two dozen of them still had eggs, and thirty-one had a parent brooding a very young chick. But Nettleship's guess that this area would have ample chicks thirteen to eighteen days old was amazingly accurate. We encountered only four nests where the chicks were older than eighteen days.

Our cases filled with chicks, we signaled for Reddick to pull in. Seas had calmed enough for him to use his powerboat, which made the pickup relatively easy. Back on land, we loaded the cases in our rental car and were in the air on Noyce's plane by midafternoon, staying below ten thousand feet in hopes of maintaining moderate temperature in the baggage compartment.

We landed in Bangor at 4:45 p.m., and I quickly hopped out to open the baggage doors, even though the proper rules would have been to stay inside until the customs agent gave the word to disembark. With enormous relief, I heard the peeping of the chicks, and I cracked the doors just enough to see that they were okay. The primitive high-low thermometer that we placed with the chicks before takeoff showed that the temperature had stayed between 55 and 72 degrees. Had the birds died, it would have likely ended Project Puffin. We later noted in the project journal, "BIRDS LIVED!"

The puffin chicks' arrival in the United States was hardly dignified. The chicks were tested for Newcastle disease by Dr. Freeman, a friendly and thorough USDA veterinarian who inserted a cotton Q-tip up each fuzzy bottom. We were surprised to discover during the inspection that somehow we had double stuffed two of the cans. This brought our total flock to sixty-eight, which was reduced to sixty-four after four were transferred to Roger Williams Park Zoo for quarantine. After the

inspection, I was free to continue on with my prize of sixty-
four chicks to Egg Rock.

The last short leg of the flight brought us safely into Wiscasset at 7:45 p.m., where we gave each chick a tiny mummi-
chog (killifish)—the only small fish that we could readily capture. We had not fed the chicks during transit to keep them
clean from excrement. (I knew that puffin chicks could sometimes go without a feeding for a day with no ill effect.)

By 8:30 that evening we were at the Bremen dock looking across at Hog Island, where two boats were waiting, our
research boat, the *Lunda,* and the Audubon Camp's appropriately named *Puffin III.* We transferred the chicks to the *Puffin III,* because it was the larger and more stable vessel. Kathy
and I boarded the *Lunda* and sped ahead to make sure that
everything was in order. It was nearly dark, but we could see
the landing site thanks to the Coleman lanterns Nina and Leslie were swinging to guide us in.

Joe rowed the chicks ashore standing tall in the Audubon
camp's wooden dory. It was a memorable sight to have the
birds arrive in the dark, with the Coleman light providing a
warm glow. By 10:00 p.m. we started putting the chicks into
burrows in ascending size, putting our smallest in burrow 1,
second smallest in 2, and so on until they were all tucked in
with each chick receiving half of a smelt.

We finished at midnight, seventeen and a half hours after
collecting the first chick. We were exhausted but exhilarated.
We toasted the chicks' arrival with a sip of Newfoundland
Screech rum.

The next morning, I resumed my role as Hog Island birdlife instructor, but my mind was 98 percent still at Egg Rock,
where Kathy and Roger were elated to find that all of the chicks

had survived the night and consumed most of their evening feeding. For their first full meal, the chicks were fed four smelt halves, each cut with a tapered end to help the pufflings swallow the food. After each meal, Kathy and Roger collected and buried leftover fish to reduce the risk that we would attract gulls; they recorded the number of leftover fish as an alert that a chick had lost its appetite. Because the fish were frozen, we continued following the recommendation of Joe Bell from the Bronx Zoo, who prescribed stuffing one of the fish with vitamin B1, vitamin E, and a multiple maintenance vitamin capsule. This vitamin-loaded fish was placed on top of the stack so that it was the most likely fish to be consumed.

Two days after the transplant, between Hog Island sessions, I went back out to Egg Rock with Tom French, the animal life instructor at Hog Island. As dusk cast its shadows and a wind began gusting past thirty miles per hour, Kathy, Tom, Roger, and I scrambled for rocks to place in front of the burrows to keep the doors tight for the night. Later, Tom dispatched thirty-five black-back chicks. This was difficult for all of us, but we were convinced that the future of the project depended on removing the great black-backed gulls from the island.

There was a brief calm before the next morning's thunderstorms. Wind and rain swept over the island, and we had to lash everything down with multiple ropes. Pools of water gathered under the tent platform and began to rise but subsided before drenching the tent. The wetness brought a cloud of mosquitoes that hounded us from dawn through much of the day.

The puffin chicks ate through it all, morning, midday, and evening. Though the dug burrows held the promise of cooler temperatures, our prototypes were soaked by the rain, so we put that idea on hold and became fully committed to making the ceramic burrows work. To reduce temperatures in the tile burrows, we set about digging larger chunks of sod to further insulate the roofs.

Severe thunderstorms and wind gusts of nearly forty miles an hour persisted into the next day, collapsing our flagpole and tent awning, allowing rain to blow through the tent's front door, the only thing between us and the driving rain. But this didn't amount to much cover at all, especially when we discovered that some of the horizontal gusts drove rain *through* the pores of the canvas.

On July 21, we took our first measurements of the chicks since placing them in the burrows. Many of the chicks had dried fish on their bills, something that seemed odd compared to wild-reared pufflings. We also worried that the open end of the eighteen-inch burrows was letting in too much light, a feature that would be abnormal in the puffin's typically dark burrow. I immediately began to cut new plywood door covers with arched entrances that would reduce the amount of light entering the burrows.

The puffin burrows on Great Island were always notable for having a straight entrance tunnel that turned left or right into a dark nest cavity. Now I began to realize that this was not just about keeping out intruding gulls; I now suspected a second function related to establishing a defined toilet area—the secret to keeping clean (and waterproof) in the often muddy burrow.

We also observed that some of the fish were being rejected, probably because they were too large. We removed the uneaten fish as we cleaned out the accumulated guano daily. Noting the condensation forming on the walls of the burrows, I worried that the ceramic burrows would not only be too hot but too moist. How simple a rock crevice now seemed. Somehow, our efforts to create simplicity were becoming increasingly complicated. Happily, through all of this, the chicks thrived.

We had hoped to feed the smallest chicks on a diet of tiny mummichogs for the first week that we had seined up from nearby Greenland Cove, but five days after the transplant, we had to bury three-fourths of our stock because it started to rot. We piled on more sod and replaced the killifish with a cache of grocery store smelt.

These fish were a little smaller than the smelt we were feeding the older chicks, but they were still eight to ten inches long. We solved that problem by feeding the little chicks just the tail end of the fish. There were, however, always challenges. Now the cut smelt were dribbling oil and guts onto the puffin chicks' breast feathers. The sinking guillemots continued to haunt me.

Despite our worries, the chicks continued to have good appetites and grow. Then, just as we were feeling some hope, we discovered that five of the pufflings were wet around the head and back, and their breasts and bellies were dirty. Nina and Roger theorized that the wetness was due to condensation from hot bird breath and excess droppings in the back end of the burrows where the chicks spent most of their time. They estimated that two-thirds of the burrows were filthy or wet. As if that were not enough, they also found ticks on the faces of two chicks.

With this news, they began a major cleaning of the tile burrows. It was an ugly task. The crud was scraped out of the corners with a spoon and then pulled out with a board. Wiping with a damp cloth left the burrows relatively clean, but it was traumatic for the birds, which were placed in a box while

Steve Kress and Kathy Blanchard inspect a translocated puffin that is being reared in a ceramic burrow. Occasional inspections were necessary to note the condition of plumage. Photo by Duryea Morton

the cleaning was under way. Desperate to reduce the accumulation of poop, we even reduced the meal size by half. We were clearly flying blind, but somehow the chicks lived.

We were also worried that extreme tides might flood our tile burrows, which were positioned on the rocks, and even stressed that somehow the chicks might stub their toes coming and going from the artificial homes. So we built wooden ramps extending out of the tile burrows for easier departure and set a six-foot-tall tide gauge at the edge of the island out of concern that a super-high tide would find its way into the tile burrows. One morning, the gauge was missing, warning us of a near disaster.

A welcome sight was the surprise appearance of Bob Noyce and a friend on their sailboat. Roger and Nina gave him a tour of the camp, ran down the mechanics of feeding, cleaning, and measuring the chicks, and pulled out a chick to show Bob how much it had grown in a week and a half. Perhaps Bob and his friend were impressed by our methods, but no sooner had they left than another tick was discovered in the nostrils of nestling 52. We worried now that we were importing not just puffins but puffin parasites that might infest future nestlings. We had inspected all of the chicks before their arrival from Newfoundland to ward against such a problem and had removed many ticks while on Great Island.

Ticks were not the only parasite of concern. A couple days later we noted that feather lice had infested puffin 57. Puffin 60 also had a few, but most were clean of lice. We consulted Christine Welch, a Damariscotta veterinarian, and she provided a powder for treating the chicks to remove the lice.

By the end of August, fledging was imminent. Weeks earlier, to test the chicks' waterproofing, we placed chick 59, which had a soiled breast, in a large tide pool. Alas, its waterproofing was poor, and it started to sink. We tried the same with puffin 58, a cleaner-looking chick; water quickly beaded up on its back and head. It was a much more promising sight. Other researchers had previously reported that puffin fledglings start coming out of their burrows for nocturnal exercising sessions in the days before fledging. This shift to a nocturnal habit is likely their way of reducing risk of gull predation. To give our pufflings a chance to exercise their wings in the open air, we built chicken-wire exercise enclosures around the burrows. These were intended to reduce the chances that the chicks would wander into neighboring burrows. But as the chicks approached fledging age, they became increasingly energetic and lost their appetite. Eventually, our makeshift exercise pens could not contain them, and they scrambled forward toward the compelling sound of the surf and the horizon.

We soon abandoned the exercise pens because we feared the chicks would injure themselves, and it seemed nothing could contain them against their instinct to reach the sea. Instead, we devised a simple activity indicator by resting a wooden dowel on a couple of nails in front of each burrow door. When the chicks stepped out, they dislodged the dowel, a clue that they had come out for an exercise session or perhaps fledged.

We continued to test the oldest chicks for waterproofing and traded the tide pool test for the "cooler test," placing the nearly fledged chick in a large ice chest filled with salt water. Most of the birds showed water nicely beading up and rolling off their backs, but many had soiled breasts, which gave us continued cause for concern.

We did not know whether soiled breast feathers would affect waterproofing, but to give them the best chance for success, we set about more burrow scrubbing and began digging more burrows into deep pockets of soil, eventually digging twenty-five burrows to replace tile burrows. We then moved some of the still-clean oldest chicks into these burrows. Nina did most of the burrow digging at first because she seemed least vulnerable to ragweed pollen, but soon she too was swelling with an allergic reaction. Groggy and dosed with medicine, she hand dug the burrows at least eighteen inches deep, curving them as puffins would naturally dig their own homes. Roger was the most allergic to the ragweed and grass pollen that now covered our clothing, and by mid-August, he would have to leave the island, ending his summer at Egg Rock. As we debated the pros and cons of moving more of the chicks from the tile burrows to newly built, hand-dug sod burrows, we noticed that puffin 58 had a cut on its bill from the chicken wire enclosure. This injury prompted the already worried team to move more of the puffin chicks out of the tile burrows and into new soil burrows.

For the coming night, however, there was a more imminent threat—a subtropical storm was heading our way. We debated whether to leave the island but decided to hunker down. We secured the tent with salvaged lobster lines and the tile burrows with huge rocks Roger hauled from the intertidal zone. The rain started near midnight, and by morning, we were recording gusts of thirty miles an hour and blowing sea spray had soaked everything inside the tent.

Roger checked the burrows at 4:30 a.m. and discovered that our first chick of the year had fledged—into the storm! Then at 8:00 a.m., Roger discovered that one of the chicks in an earthen burrow had escaped by digging around its hardware cloth door. The bird had dug so hard into the burrow's foundation that the structure caved in on its own a few hours later. That would have all been fine except that the chick left before we could band it.

On August 7, a research team from the Mount Desert Island Biological Lab in Bar Harbor paid us a visit. Ten of our chicks were part of a study of the effects of DDT on nasal salt gland function. The USDA vet who greeted us at the Bangor airport, Dr. Freeman, was also along for another inspection of the chicks for Newcastle disease. The Mount Desert Island lab group took blood samples from the ten birds that were part of the DDT study, and Freeman probed the cloaca of the remaining puffins.

Five of the chicks were given doses of DDE (a derivative

of DDT) suspended in vegetable oil, while the five control birds received the vegetable oil only. Although DDT had been banned just two years previously in the United States, the chemical is long-lasting and degrades to DDE, which is still highly toxic. We struggled with the knowledge that some of the chicks would be sacrificed for science, but this was tempered by knowing that their fate would help build a solid case against DDT.

The next day brought very distinguished guests, retired Audubon president Carl W. Buchheister and his wife, Harriet. We showed them a puffling, which Buchheister photographed with much delight. I was personally indebted to Buchheister, for it was his stories about his journeys to Matinicus Rock that made me dream of one day living among seabirds. He had also created the student assistant program that brought me into the Audubon fold. This moment was certainly a full-circle visit, with Carl and Harriet my guests on Eastern Egg Rock and me showing them a puffin.

On the way back to Hog Island from Egg Rock, Carl stood next to me like Washington crossing the Delaware, upright, eyes forward. Tiny Harriet sat on the seat behind me, bundled up with many blankets, looking at the froth created by the outboard. Suddenly, I heard a high-pitched squeal, "Storm-petrel!" Somehow, Harriet spotted the tiny robin-sized bird from under her bundle of blankets. Carl and I turned to see the bird at a distance and marveled at her vision as the dark speck soon disappeared.

More weather challenges came on August 12, when our burrow thermometer recorded 90 degrees inside one of the ceramic burrows. Our worry that the ceramic burrows would turn into ovens was proving well founded. We tried to convince ourselves that a chick was sitting on the thermometer, but that was unlikely. All we could do was pile more sod on top of the tiles for better insulation.

On August 14 I brought the Mount Desert lab team back out to the island to observe their ten chicks. After they left, Roger opened all the remaining burrows, except for those holding the ten Mount Desert chicks.

At summer's end, gull numbers were increasing just as our chicks were fledging. We alerted Frank Gramlich, and he quickly arranged for a return visit. This time he carried twenty pounds of Starlicide-poisoned fish. He and Kathy walked around the island for two hours, Gramlich scattered the poisoned fish while Kathy dropped unpoisoned fish in the hope that the gulls would be fooled. Apparently, even though Gramlich had not visited the island since spring, the gulls remembered him and avoided the poisoned bait. Yet most of the unpoisoned bait was consumed.

It would have been remarkable enough that the gulls had remembered their kin dying from the poisoned bread, but to generalize this to fish! We pondered this for some time and came to suspect that the gulls remembered Gramlich from his earlier visits and associated him with danger. Research with crows, another intelligent bird, has shown that crows can recognize individual people by facial features.[2] Apparently, some recognized Gramlich and warned others to stay away from his "gifts." It is also remarkable that they readily ate the fish that Kathy dropped. What was certain was our new insight about gull intelligence and that we should never underestimate this most successful predator.

Fall shorebird migration was in full flight, and our mid-August walks on the edges of Eastern Egg stirred up scores of ruddy turnstones newly arrived from the Arctic. That night, Roger said Jupiter was shining on the water when seven more fledglings departed. The birds that remained were now appearing to exchange burrows indiscriminately. A couple of times, two chicks were seen in the same burrow. On August 16, three dead immature black-backs floated ashore, evidence that at least a few of the younger, less experienced gulls had fallen for Gramlich's bait.

Two more chicks left on August 17, and three hummingbirds buzzed the island on their way south. Four more pufflings left on August 18, despite sustained winds of twenty-five miles per hour that left the remaining seventeen birds wet. Six more left before sunrise on the nineteenth, five more on the twentieth, and three more on the twenty-first of August. Then we were down to three of our own pufflings and the ten Mount Desert chicks, which the lab picked up that day.

Finally, by dawn on August 24, the final three fledglings had departed into the Atlantic. We celebrated our success: the departure of the last of our fifty-four chicks. None of the Newfoundland transplants had died either in transit, despite the risk of overheating in the plane hold, or on the island. And in spite of our primitive burrows, all the pufflings fledged with normal body weights. Although we were still not sure if their waterproofing was adequate for survival, we had achieved a 100 percent fledging rate. We had learned an enormous amount about rearing puffin chicks and were encouraged by their natural vigor and adaptability to our experiments with food, vitamins, and burrows.

On September 3, I wrote to David Nettleship to report that all fifty-four puffin chicks fledged from Eastern Egg Rock. I hoped he would see that the chicks were in good hands and that we could continue with even more puffins the next summer. I was filled with hope that we had taken a huge step toward our ultimate goal of reviving this lost puffin colony.

6

Soaked Sod and Puffin Condos

I made my first trip of the season to Egg Rock on June 18, 1975, with Tom French and two new research assistants, Kevin Bell and Mike Haramis. This new team was necessary because Nina and Roger had made other plans and Kathy had accepted a position in Canada to help start a seabird program with the Quebec Labrador Foundation. In their place, I was fortunate to land a dream team with exceptional experience. I had met Mike Haramis, a fellow grad student majoring in natural resources at Cornell, the previous year. With his duck hunting experience and career focus on wildlife management, he was perfect for puffin work and gull control. Likewise, Kevin Bell, son of Joe Bell, curator of birds for the Bronx Zoo, was a natural for helping me rear puffin chicks.

Our first view of Egg Rock was the same dispiriting scene that perturbed Frank Gramlich the year before—four hundred gulls still dominated the island. Gramlich had visited a few days earlier to do his work, so there were about forty black-backed and six herring gull carcasses to collect. We crushed thirty more eggs and dispatched about sixty black-backed chicks. As distasteful as this job was, we saw plenty of reminders why this was necessary. Here and there in the grass and boulders were the remains of gull predation, including left over body parts of cormorants, storm-petrels, guillemots, and eider ducks.

That we had nurtured fifty-nine out of sixty puffin chicks to fledging age during the past two summers was enough for David Nettleship to deem "quite acceptable" my subsequent request to expand collection of chicks from Great Island to one hundred in each of the next three years. But as the last two summers also proved, we were a long way from perfecting human-made burrows. The extremes in temperature in our tile contraptions and drainage problems in our sod homes had been on my mind all winter. Soon after arriving on the island, we set about trying to upgrade our architecture.

We walked to a knoll where we hoped to find suitable soil and plunged our soil auger in to measure depth. Usually, we heard the clink of granite just inches under otherwise promising sod, but persistent testing eventually helped us discover one hundred deep pockets of soil suitable for digging with small garden trowels. We dug half an arm's length in and then twisted the tunnel sharply to the right or left, often dodging a buried rock, to create a nest chamber. The L-shape would provide both natural darkness and protection for the chick, but it greatly limited the digging options. After many aborted attempts due to shallow soil and hidden rocks, we tunneled ninety-nine burrows in our first three days. Then we pulled out the ragweed that grew near the new burrows. It was only about two inches high, but we knew it would shoot up quickly to a height of three feet or more, creating a tangle of obstacles for departing puffin chicks and clouds of debilitating pollen. We didn't want to lose anyone as we had the previous summer, when Roger had to leave early because of his severe allergic reaction to ragweed pollen.

On July 1, Kevin, Mike, and I set out from Hog Island to

Eastern Egg to set up permanent camp for the summer. When we arrived, we found six-foot swells rolling into the landing ledge. Even in calm weather, the landing was tricky, because we needed to unload fifty-pound water jugs, duffle bags, and masses of gear wrapped in plastic bags from the *Lunda* into our plastic skiff, the *Geezer*. I managed to row into a narrow channel, where Mike hopped out. I quickly threw the gear onto the shore before rowing back to the *Lunda* for the next load, which was now tossing wildly on the mooring. Kevin passed me another load, and I managed to get it ashore to Mike, who clung to the slippery rocks.

The third trip was different. Kevin climbed into the *Geezer*, which was half filled with water jugs and gear. We managed to unload the gear but we were immediately sucked deep into a guzzle by a receding swell. In an instant, we were looking up eight feet at Mike and the crest of the next curler that poured into the *Geezer*, flipping the boat and us into the frigid surf. Gasping and stunned by the frigid water, I held onto a lifeline from the skiff while Kevin clung to the rocks. We were hit by yet another wave before Mike reached down to pull us out of the water. We were lucky that all Kevin lost was a new boot and the only injury was a barnacle-scraped finger.

It was a shaky start to the summer and yet another reminder of our vulnerability on this outermost island. While preparing to receive the next class of puffin chicks, we kept busy by netting and banding Leach's storm-petrels to learn how many frequented the island. Grouped in the same taxonomic order as albatross, these robin-sized seabirds fluttered ashore under cover of darkness with surprising dependability at about 10:00 p.m. every night. Hundreds circled over and many landed on the ground to make their way through dense grass to underground burrows. Especially on dark, moonless nights when the gulls could not see, the storm-petrels arrived on masse, chuckling like so many avian demons while their musical purring tumbled from soil and rock crevice burrows.

Tom French was particularly enamored with the storm-petrels. Over the next few years, while waiting for puffins to come back, Tom and other helpers on Eastern Egg Rock would band about two thousand storm-petrels. Of these, Tom personally banded fifteen hundred. Occasionally we recaptured storm-petrels that had been previously banded on Kent Island in the Bay of Fundy, Bon Portage Island, Nova Scotia, and even one from Iceland. Tom's enthusiasm came from his boyhood love of the chirping of birds in the Atlanta suburbs and the adventure and sense of fearlessness he found in his local Boy Scout troop. One year he went with the elite hikers in his troop to New Mexico for ten nights of wilderness backpacking at Philmont Scout Ranch. Trekking among rattlesnakes, bears, and mountain lions at altitudes that could reach twelve thousand feet, he and the other teens in his crew had to make all the decisions where to go. Tom's background so impressed the Atlanta Bird Club, the city's forerunner to its Audubon chapter, that he won its first-ever scholarship to be a camper for a two-week session on Hog Island in 1972. At the time, he was majoring in biology at Georgia State University.

Sometimes Tom became a little too enthusiastic. One day on Franklin Island he broke his thumb trying to catch a storm-petrel with his bare hands. He did not have a functioning radio connection and had to wait two days gritting out the pain. Later he learned that he would need to be hospitalized for a couple of days to manage the resulting infection.

Kevin Bell had the perfect background to help raise puffins. He grew up living with the animals at the Bronx Zoo. In 1958, when Kevin was five, his father, Joe, moved the family inside the zoo, to quarters behind the reptile house. Whereas dawn for most New York boys arrived along with the first honking of horns and the screeching of subway brakes, Kevin's day began with birds squawking, tigers growling, lions roaring, and primates grunting. At age ten he was doing chores for his dad, everything from cleaning dung to reaching into the incubator to turn over eggs and look over the hatchlings at night.

This life got the attention of the media, with one newspaper headline declaring him to be "The Boy with 2,830 Pets." A young Mike Wallace featured him on a local news segment as the little boy who lived in a zoo. Kevin would say that the highlight of the interview was riding in the Bronx in Wallace's convertible and being paid a hundred dollars.

It was Kevin's dad, Joe, whom Kathy called when we needed advice on puffin chick diets, because at that time the Bronx Zoo was the only zoo with experience rearing puffins. In September 1970, it was the site of the first successful zoo hatching of a tufted puffin. With his instincts and passion for caring about wildlife, Kevin proved to be a major asset as we fumbled our way through methods of rearing puffin chicks.

Bob Noyce could not fly us to Newfoundland in 1975, but he agreed to pay for the charter flight. We flew on July 12, piloted this time by Jack Thompson and his wife, Nancy, in their Beech Queen Air. We were not going out to the island until July 14, so we took a drive down to the shore looking across at Great Island, but soon found ourselves picking through the bodies of puffins, murres, and razorbills discarded on the gravel beach, victims of local fishing nets. Some of the birds were skinned; others were completely intact.

On the fourteenth, we awoke at 3:00 a.m. and met up with David Nettleship to boat over to Great Island. We immediately set to our work probing puffin burrows to find the ideal size chicks.

We collected our quota of a hundred chicks in about three hours by focusing on the steep, grassy slopes where the soil was relatively shallow, removing ticks from the chicks and ourselves when we saw them. We were fortunate to have perfect weather all the way into Bangor, where once again the USDA vet probed the chicks for Newcastle disease, this time passing six of the chicks to the Roger Williams Park Zoo for quarantine. We continued on to Wiscasset and then drove the final leg to the end of Keene Neck Road, where the *Lunda* and *Puffin III* awaited. We entrusted the puffin chicks to Joe Johansen aboard the *Puffin III* while I sped ahead with Kevin, Mike, and Tom in the *Lunda*. This year we plowed through heavy seas in the near dark to get out to Egg Rock. By the time the chicks were ashore, tucked into their burrows, and given six smelt halves, it was 2:38 a.m., fourteen hours since leaving Great Island. We were exhausted.

We checked all burrows in the morning and found only six discarded smelt halves. A week after their transit from Great Island, all ninety-four chicks were thriving. In their first week, most had doubled their weight. We noticed, however, that the chicks seemed to be learning to recognize our approach with meals, and many started to chirp with anticipation for their next feeding. This was disturbing in that we did not want the chicks to learn anything from humans and become less than wild after fledging. In response, we set a no-talking rule when we were near the chicks.

We began banding on July 26, removing screens from the burrows housing the ten biggest pufflings. Things were so uneventful that the next day Mike and Tom took in the sunset with cups of coffee. "About 100 yards off shore," Tom wrote, "we spotted a small baleen whale coming out of the bay and heading due south. It surfaced about every 30 seconds." Kestrels soared overhead, nighthawks cut the air at dusk, seals basked at low tide, and a large shark circled the island, showing off a huge dorsal fin. One day, an albino barn swallow spent a morning on the island. These seven acres were as productive for bird-watching as a great forest yet offered the fruitful marine life of a whale-watching trip.

In this lull between routines and crises we took turns supplementing our canned, dried, and packaged foods by picking lettuce and beans out of the garden that Tom had planted in June. Sometimes Kevin, Tom, and Mike would collect periwinkles (invasive snails) at low tide, steam them, and sprinkle them over their salad. "Escargots," Kevin said. Some of nature's bounty did lack aesthetics. Dogfish had this peculiar habit of curling up in the frying pan "like rubber," said Tom. Lobstermen had got wind that Eastern Egg Rock was now inhabited in the summer, so they would sometimes sell us a few discounted bugs. Better yet, on a few occasions, a storm blew a loaded lobster trap up onto our rocks.

At the end of July, the chicks were eating so fast that we knew they were fattening up for fledging time. Hundreds of gulls were now loafing around the shores of the island, waiting to snap up our precious chicks on their way to the sea, so on July 30, I brought Frank Gramlich back out, along with Carl Buchheister. Frank put out another buffet of poisoned bread, then shot six adults and left the bodies of three on the rocks to signal other adults that this was a dangerous area.

Buchheister came along on the trip to see the fledglings, and he helped me check the burrows. Though not a scientist by training, he was a keen observer, and I could see that he truly believed success would eventually come my way. Together, we discovered that a couple of pufflings had started coming out at night, so we immediately accelerated our banding of the chicks with Buchheister lending a hand. We surmised that a couple of pufflings had fledged, only to find them hiding under nearby rocks. Later we realized that they had fled burrows that flooded.

August 5 brought heavy rain, but most of the burrows were still okay. We were very wary of our hand-dug burrows at this point, fearful that the next soaker would flood the more vulnerable ones. We were so close to fledging time—if only the rain would hold off for another week, then most of the chicks could fledge.

Our optimism was dashed on August 8, when heavy rain during the night and morning flooded twenty-two of our hand-dug burrows. When the rains subsided, we found several burrows completely filled with water. Somehow the birds were still alive, but most were soiled on their breasts, which made us nervous so close to fledging.

We brought the most soiled ones back to the tent, where the chicks were dried and then turned loose, the tent now serving as a puffin holding pen. From this near disaster came one of our greatest innovations: "puffin condos"! The new burrow design evolved from the use of sod. Like prairie farmers who made their homes from sod on the tall-grass prairie, we found Egg Rock sod to be ideal for construction and set about building L-shaped burrows in long rows. The L-shape provided a long entrance tunnel with a bend into the nest chamber, where the chick could find a snug hiding place and safety from the long neck of a probing gull. The sod burrows were built on top of existing sod, covered with vinyl-coated welded wire and supported by wooden lath, and then topped by another layer

After hand-dug burrows flooded, we cut sod and laid it out in L-shaped formations to create "puffin condos" above ground. This photo shows burrows under construction with the first floor of the burrows in the foreground. A finished row of burrows with sod roofs forms a second row behind. Photo by Stephen W. Kress

of sod for roofing. Each burrow received a brightly painted yellow placard with black numerals for a street address.

Despite the flooded burrows, the chicks' overall condition was very encouraging. Although we had lost one puffling, ninety-three were still eating, and the condition of their feathers was much better than the previous year, when the chicks had been reared in the ceramic burrows.

That night, the chicks began their march to the sea. Thirty-three pufflings knocked over the wooden dowels that we used to indicate nocturnal activity and then returned to their burrows by morning. Two chicks kept going, becoming the first to make their way to the ocean.

By August 13, more than half of the chicks were disturbing their low-budget activity indicators and six had fledged. The next night, sixty-two of the remaining eighty-five chicks ventured outside. In the wee hours of August 15 and 16, Mike and Kevin hunkered down among boulders along the shore, listening for the metallic, tinkling sound of metal bands against granite. Always observant, Kevin noted that the chicks were more active whenever clouds covered the moon. If this were true, it would be an interesting way to avoid predation from great black-backed gulls, which sometimes feed at night.

Mike and Kevin observed five chicks on their trip to the ocean . . . all but a single puffin moved by foot and took short, quick flights only to clear obstructions in their path. A single puffin was a strong enough flier to reach the ocean from the granite ledge to the northwest of the puffin colony. All chicks moved to the sea between 1:15 and 3:15 a.m.

I arrived to witness the procession on August 17 and, with Kevin, watched three chicks fledge. At least thirty-five fledged during the next three nights. I could finally relax, knowing that we had done everything possible to send them off in top condition. The puffins would not set foot on land again for at least two years. I just hoped their eventual homecoming would be on Eastern Egg Rock.

One puffling created quite a stir the next morning when Audubon campers headed to Egg Rock on a seabird-watching field trip. I was leading the trip and was thrilled to discover the puffin fledgling about a mile from the rock. The thick fog probably prevented it from flying far from the island, and it was so tame that we were able to approach within about a hundred yards. At the time, I worried that it was overly tame around people but have since seen wild-reared fledglings behave in a similar manner. And most important—this fledgling was completely waterproof.

By the end of the fledging time, more chicks had fledged

than the previous two years combined. To celebrate the end of a very successful summer, Kevin and I brought out lobster and wine from the mainland and on a whim went eight miles farther offshore to Monhegan Island, where I landed a three-and-a-half-foot hake. Tom wrote of our last meal together as a team, "We fried fish to go with the wine, lobster, salad; donuts and coffee for dessert. What a meal!"

In his final note for the summer, Tom wrote, "All kinds of things were eaten including 14 pounds of peanut butter, one and a half gallons of strawberry preserves, pollack, lobster, cod, hake, shark, mussels, periwinkles and a hell of a lot of freeze-dried stuff."

As for the puffin chicks, they ate about twenty thousand smelt and ten thousand vitamins. There were only three casualties. All the others left healthy and strong and, we hoped, sufficiently waterproofed. Yet we still had no idea if they left thinking that Eastern Egg was their home or how long it would take for them to return.

Splat. Gull crap in my eye! It was somehow fitting on this first day on Egg Rock, mid-June 1976. Winds were so stiff that the usual thirty-minute ride from Hog Island to the rock took an hour and a half. Then it took five trips to shuttle all the camping gear and food ashore in our patched-up *Geezer*. I had come to understand that if rain and fog were a nuisance for boating, wind was a greater threat (especially if from the north), blowing into our landing ledge. Fortunately, this was a south wind. And that is why it took so long for us to reach the island. Once at Egg Rock, we had enough lee to get ashore, but putting up the tent in thirty-mile-an-hour gusts was difficult. The south wind howled all night long. Proof of nature's resilience to these gales, we could still hear storm petrels arriving in the darkness, returning to their nests.

I was always wary about leaving the *Lunda* on an anchor for long, but I had confidence in the huge block of granite that Joe Johansen had rigged for a mooring. Thus I was all the more shocked the next morning when I went to the landing to check on the *Lunda,* and found her missing! In a panic, I searched the horizon and at first found no trace of the boat. Then I spied a white speck against a forbidding outcrop of basalt known as Old Hump Ledge about two miles north.

Robert Wesley, a friend from Ithaca whom I had recruited as a research assistant, gave it only a moment of consideration before untying the *Geezer,* and together we quickly launched it to retrieve the *Lunda.* Robert had grown up in upstate New York, far from the sea, so he took this all as "must be normal" behavior for those that live along the coast. For my part, I was worried that the *Lunda* would be smashed against Old Hump Ledge or drift out of sight completely. Fortunately, the wind was still blowing from the south, so we made great time rowing our little orange tub much farther than I'm sure its designers ever imagined. Dur Morton's caution, "Don't let the puffins interrupt the Audubon camp," was on my mind as we rowed. If I didn't retrieve the *Lunda,* the camp would have to waste good time, fuel, and Dur's patience to fetch us off Egg Rock.

Apparently, the *Lunda*'s mooring had been set in water that was too deep, and the heavy seas had lifted everything off the bottom. Both the *Lunda* and the mooring had floated free and gotten hung up at Old Hump Ledge. Fortunately, the mooring had snagged on a ledge, preventing the boat from smashing on Old Hump. We happily boarded the breakaway boat and were relieved to head back to Hog Island. My reputation as a boatman was less than stellar in these early days, and such narrow scrapes were hard to outlive. I had taken a coastguard auxiliary boat safety course on Cayuga Lake, but the details of lake boating did not transfer well to the Maine coast.

After recovering from this mishap, I returned to Egg Rock on June 22. Dealing with a resurgence of great black-backed gulls was a priority, for we found that 230 adult gulls had reclaimed the island, sitting on all of the highest promontories from which they defended their nests among the nearby clumps of grass and weeds. Eagle Scout Tom French was back for a second summer and brought out his shotgun.

Tom stood at the tent and downed a pair of black-backs with a single shot. Instantly, all the gulls on the island took off from the rocks and flocked to the spot where the two dead birds lay. After hovering overhead and sitting nearby, the flying birds all took a good close look, but they didn't seem to be afraid of us. Tom could have easily shot some of the gulls out of the air, but guillemots and eiders also flew up when the blast sounded. We hoped that the eiders and guillemots were adapting to the occasional shooting while the gulls would learn to become more wary as they learned that they were the intended targets.

Mike "Zac" Zaccardi, an adventurous educator from Glen Helen, joined the team and, through his own experience, confirmed Tom's observation that shooting a few gulls would result in an exodus of other gulls from the island. With each passing day, we could see that the gulls were learning to avoid people, which meant that they would need to go elsewhere to nest successfully. No one liked shooting gulls, but we were convinced that the gulls would undo all of our efforts for puffin restoration. Success meant breaking up the gull colony.

In between the gull assault, I planted sea lavender, arrowgrass, and seaside plantain that I had gathered from a village campground in the coastal fishing village of Jonesport. Later I would transplant roseroot sedum from nearby Matinicus Rock. These native succulents were conspicuously absent at Eastern Egg Rock, and my thinking about bringing them back to the island was similar to my rationale for puffin restoration. People had caused the plants to disappear, and therefore it was appropriate for others to help bring them back. Unlike the puffins, however, which were well-documented former residents of Egg Rock, there were no early botanical studies. I wondered if these island specialties had been eaten long ago by the sheep brought to the island for grazing.

Sheep grazing continues today on a few Maine islands, but it was once widespread and likely occurred at Egg Rock in the 1800s, when most islands were used as island farms. Nearby island names such as Hog, Cow, and Ram are reminders of this agricultural history, which dates back to the early 1700s on the Maine coast. Small islands like the Egg Rocks were favored for grazing sheep because they offered natural isolation from such mainland predators as bear, wolf, and cougar.

Sheep likely gobbled up the last of certain native plants, and they may also account for the presence of such invasive species as timothy and quack grass, which may have found their way to Maine islands attached to sheep wool. Some of these grasses may even have been planted on islands to improve the grazing. These tall grasses reduced the amount of open space required for tern nesting. Now, in an ironic turn of events, we were using sod from the invasive grasses to build puffin condominiums.

We brought other plants as well to our island home, which now supported a small garden of tomatoes, lettuce, green beans,

squash, zucchini, and radishes. Robert had an affinity for gar-
dening, but all plants were of interest to him, and he was already the go-to guy around Ithaca for native plant identifi-
cations. He was impossible to stump, even with grasses and
sedges. Tall, lanky, and peering out from behind thick glasses,
he was an unlikely ladies' man, but plants and girls were his
two favorite topics and women of a certain type did admire his
botanical knowledge. Isolated on Egg Rock from female company, and with plenty of time on his hands, Robert was happy
to jump into a botanical study documenting the variety of
plants on Egg Rock. He conducted a detailed transect study in
which he counted and identified the stems of individual plants
in alternating meter squares from the high-tide line to the interior of the island. This survey proved useful in later years,
helping us understand the impact of our bird restoration on
island vegetation.

We have since learned that the gulls were very much
affecting the island vegetation, disturbing the soil by pulling
plants for nest building and trampling down the otherwise tall
plants near their nests. Their excrement fertilized the soil, but
it could not compare to the flood of avian guano that would
result from our future success of bringing back thousands of
terns and laughing gulls.

With memories of flooded burrows the previous summer,
we set about replacing all of the hand-dug earthen burrows.
Tom was now my veteran partner, and he led the new construction effort.

Such work was a reminder of the isolated, odd world we
had chosen. It was an era where a teenage Bill Gates and Paul
Allen were starting Microsoft. This was the summer NASA
landed the Viking probe on Mars to send back the first photo-

graphs directly from the red planet's surface. Motorola and Bell Labs were racing to create the first mobile phones.

On Eastern Egg Rock, our needs were simple: our tools of innovation were wire, sod, and rock, but in our own way, we kept stumbling on improvements. In digging up sod for the roofs of the burrows, Zac developed a new tool that we dubbed a sod car (pronounced "saad-caa"). This consisted of two two-by-fours separated by a three-by-four-foot plywood board. The primitive tool proved key to moving mountains of heavy turf. Some of the larger sod blocks were three feet by four feet across and about a foot thick. Moving chunks of sod this size required two strong backs to carry each load.

We had chosen the site for the sod puffin burrows to be near the tent camp so that our presence would deter predatory gulls. But the deepest sod deposits were far from the burrow construction site, requiring us to trudge with our heavy burden nearly halfway down the length of the island, a distance equal to half of a football field.

After the blocks were carried to the burrow construction site, we used machetes to carve them into walls, which were laid out in an L-shaped arrangement, so that each chick would have a dark nest chamber. Finally, the largest and thickest sod blocks were hefted into place for roofing. It was Project Puffin's equivalent to building the great pyramids.

Now we were in the fourth summer of the project, busier than ever, but still no returning puffins. Occasionally, while hauling sod and working so hard, questions about the eventual outcome of the project would creep into our conversation. Whenever there was an inkling of doubt, I would remind all that puffins spend the first several years at sea and may not breed until five or six years old. I tried to say this with confidence, but at the time there was little proof of the accuracy of these estimates. Likewise, little was known about what puffins do in their first few years. We speculated how puffins spent their days at sea and why they waited so long to nest.

Likewise, we did not know how long puffins live (we have since learned that the oldest birds can survive to their midthirties, and twenty years is common). Later, our studies and those of others would document that puffins would often return at two or three years of age. Had we been aware of these details, we might have been more worried that we had not re-sighted any of our returning birds by 1976. The window of two years had passed for the 1973 quintet, though of course it was too tiny a group to hold out any realistic hope for returns.

The National Audubon Society signaled its optimism for Project Puffin that summer by naming Eastern Egg Rock the Allan D. Cruickshank Wildlife Sanctuary. Cruickshank, an original member of the Bronx County Bird Club, was a friend of my mentor Irv Kassoy and one of the first two bird instruc-tors at the Hog Island Audubon Camp, which opened in 1936. Roger Tory Peterson was the other first instructor, and he was with us on this special day for the dedication. Cruickshank was a traveling lecturer for Audubon through the 1960s and died in 1974 at the age of sixty-seven.

There was a silent irony involved, given my purpose here. During his time as instructor, all Cruickshank could show campers on Eastern Egg were gulls, along with the occasional storm-petrel or guillemot. He did on rare occasion see a puffin in Muscongus Bay, but it was unclear what majesty the bird held for him. On one hand, he once described the puffin as a favorite among many bird watchers for its dignified upright posture. But in a small bird identification guide for Lincoln County, Maine, he also described the bird as "grotesque," because of its outsized bill. Only once did our paths cross—in 1972 when he received the Cornell Lab of Ornithology's Arthur Allen Award. I introduced myself among the well-wishers, but there was not much to say at that point about puffins. Many years later, his charming wife, Helen, presented me with a wedding present they had received from famed artist Francis Lee Jaques—a painting of a Bonaventure Island puffin.

The timing for the dedication was also ironic because that very morning, a local reliable fisherman radioed to us that he saw two puffins on the water one mile from the rock. I was wary of announcing premature success and did not mention this to the honored guests. Yet I wondered to myself if a pair of puffins had come in some way to honor Cruickshank and the concept of a puffin sanctuary.

On July 18 it was time to fly back to Newfoundland for the next puffin chick roundup. Unlike prior trips, seas were so flat that John Reddick was able to land Tom, Robert, and me on Great Island a day and a half before the actual collecting. This gave us the chance to explore more of the island and better appreciate this grand landscape and the multitudes of seabirds. I was once again dazzled by the enormity of the scene, with its splendor of great cliffs, tens of thousands of puffins, and the deafening cacophony of screaming murres and kittiwakes.

The history of diminishment and the recovery of birds that escaped complete extinction was all here. David Nettleship estimated that there were about 2.5 million breeding seabirds from Newfoundland and Labrador down to the Gulf of Maine. The enormity of the populations here helped to put David's initial dismissal of my proposal in perspective. In this grand landscape with so many birds, I could see why the idea of restoring a few birds to the edge of the range probably seemed inconsequential.

Yet even in this setting of remarkable abundance, there were many threats—some invisible to the casual eye. For example, in a 1987 paper, Nettleship and others cited how gannets recovered abundantly from nineteenth-century slaughter in the Gulf of Saint Lawrence but were again in decline because of reduced fertility from DDT.[1] Similarly, Leach's storm-petrel numbers were dropping because of dogs and cats and human development. In the 1970s, murres, razorbills, and Atlantic puffins appeared to be in a long-term decline.

Even before today's worries about climate change and warmer oceans, Nettleship identified threats that rivaled the reckless plunder of the early Canadian exploiters. He cautioned especially about oil drilling and marine mining along with other forms of industrial expansion. He was especially concerned about the precipitous decline of Canada's Atlantic cod stocks, which in turn induced fishing fleets to settle for smaller fish, such as capelin, the most important food of puffins, other seabirds, and marine mammals.

Although millions of puffins nest on the islands of northern Europe, Nettleship said that as far as Canada was concerned, the puffin was at grave risk because 72 percent of its North American population nests at a single location: Witless Bay. That made it all the more meaningful when, after I reported to Nettleship that we fledged ninety-one of ninety-four puffin chicks the previous year, he wrote back to congratulate me on running an efficient operation. Yet I knew that we were inventing chick-rearing methods as we went and it seemed that our success was precarious at best, always vulnerable to the next deluge or sneak attack from aggressive gulls.

With David's encouragement and assistance, we collected 105 chicks in 1976, giving up 5 to USDA when we returned through Bangor for quarantine and observation at the Providence Zoo. On arriving at Hog Island, we showed off one chick to excited campers before speeding off to Eastern Egg.

But the 1976 transplant did not get off to a good start. Within the first two days, one chick stopped eating and died. Another disappeared, having dug its way out of the burrow. We searched everywhere but never found it.

No matter how many black-backed gulls we shot, there always seemed to be more. Herring gulls were scarce, and one day we received some insight why. Tom saw one black-back grab a herring gull chick. It snatched the chick by the neck and started shaking it. The chick screamed in distress, which brought its parents. They hit the black-back solidly several times, but this was not enough to make it drop the chick. The desperate parents gained altitude and then dove again, hitting the black-back with enough force to make it drop the chick, which promptly escaped. More often, however, the big black-backs were the winners.

Meanwhile, Zac and Tom discovered a fresh sign of growing diversity on Egg Rock, a spotted sandpiper chick in a rock crevice and an adult writhing in fake distress to attract attention away from the chick. We interpreted the discovery as a hopeful sign that some new species would colonize the island simply by our causing the gull population to back off.

As summer progressed, we saw more migrants. One day, a group of eighteen loons floated past the shoreline, and nighthawks winged by at dusk. A yellow-and-black-streaked prairie warbler joined the crew for several days at breakfast, perching on our tent. Drab grayish yellow Tennessee warblers and flamboyant black-and-orange American redstarts snatched insects from our garden. Our transplanted sedums were also thriving, bolstered by some additional plants collected at Matinicus Rock by Harriet Buchheister. She also sent seaside bluebells, though I learned later that they were not native, and it was probably just as well that they withered.

While banding storm-petrels, we noticed that some made loud, purring sounds from their occupied nesting burrows while those in flight made a very different chuckling sound. I recorded the calls of several that were calling from under the rocks and some that called from the air with a tape cassette recorder positioned under our mist net. The underground burrow call proved especially attractive for luring petrels into the net. Nearly every time I played the tape, one or two storm-petrels would hit the net about three feet over the recorder. On our first night of netting, we captured forty-four storm-petrels, including three banded on Egg Rock the previous year. I didn't realize it at the time, but use of the playbacks for banding would lead to the use of audio recordings for starting seabird colonies.

Then, in the second week of August, just when we expected puffin chicks to fledge en masse, warnings for Hurricane Belle were issued. Tom was on Egg Rock with Bob, and I was at Hog Island. With the safety of Tom and Bob at stake, I radioed to alert them of the situation. This was the first hurricane warning since we had started the project. They had heard nothing of the pending weather. I asked if they had ample food and water to stay a few extra days without a supply run.

Tom said, "We have tubs of peanut butter and jelly. You could last for a year on that stuff. We could get crabs and mussels if we had to."

I asked if he had some extra line to tie things down.

Tom said, "Why?"

"There's a possibility we might have a storm."

"Great, we can handle that."

"Well, it's actually a hurricane."

I didn't realize the psychic trauma I had just inflicted on Tom. In 1969 he had seen the aftermath of Hurricane Camille in Biloxi, Mississippi, in particular one city block that was entirely blown away and all that remained were foundations, concrete front steps leading up to the sky, and tree stumps.

"Oh, shit! Hurricane, that'll wipe everything here off the face of the earth!" he warned.

The storm was supposed to arrive shortly before midnight. The coast guard advised me that, once the storm arrived, they would not be coming out for rescues. I explained to Tom and Bob that our own salty Joe Johansen believed that Egg Rock would probably get completely "sprayed over" but not deluged with "green water," Johansen lingo for waves washing over the entire island. I offered to come and get them right away.

Tom said they would stay.

But a new panic set in. Storm-driven waves might flood the burrows. And only about a quarter of the puffin chicks had been banded. If the unbanded birds washed away, survived, and ever returned years later, there would be no way to prove that they originated in our experiment.

So after tying down their tent, Tom and Bob drew up a contingency plan to rescue the chicks from their sod burrows if waters approached flood level. After the storm passed, Tom explained their plan. "We had four five-gallon plastic buckets. If it came to it, we were going to run around and pull the chicks out and fill the buckets with chicks and let them have the run of the tent during the storm. It would be like a chicken house." Nearly forty years later, Tom recalled, "We didn't sleep that night because we didn't want to get surprised."

Fortunately, Hurricane Belle petered out after hitting Long Island, New York, on August 10 as a category 1 hurricane. It caused a hundred million dollars in damage in New York, but by the time it got to northern New England, the winds had diminished to "only" seventy miles per hour. This was still a lot for an island of seven acres and a high point just seventeen feet above high tide.

"We stayed out the entire time," Tom said. "It was too irresistible not to be out in the excitement. The spray was pounding our skin. The ocean looked like an agitator washing machine. There was no pattern to the water. It was just big ups and downs of water everywhere, randomly agitated."

The storm was strong enough that Tom saw the two-ton navigational bell buoy that floated off the north end of Egg Rock momentarily come completely out of the water, dangling upward off its mooring chain. "You would think the monster boulders on Egg Rock are so big they couldn't move," Tom said. "I saw one come out of the water airborne. After the storm we had two or three that were lifted out of the subtidal zone—but were now high and dry—seaweed still in place. It was as if they just popped out of the water."

There was one benefit from the storm. Lobster pots blew up on shore with more than Tom and Bob could eat. They liberated most of the lobsters back into the ocean.

As for the puffins, it was as if nothing had happened. We were so proud of our new burrow design. The sod had stayed in place, and despite heavy rain, none of the chicks were wet. Two days later several puffin chicks pranced around the openings of their burrows in broad daylight. Some of the chicks sensed when we were coming to feed them and came out to the entrance expectantly for fish. The nightly procession to the sea grew over the next several nights without major incident, though some puffins needed a little help.

One evening as Tom and John Hunt, a marine life instructor from Hog Island, were examining a flatworm they found in a tide pool, they heard a bird climbing the base of the tent. "I ran out and grabbed what I was sure would be a storm-petrel but to my great surprise—it was puffin 37!" They noted that the bird was a fledgling with just a hint of down left on its neck and was probably attracted to a light inside the tent. They put the bird back in its burrow.

Then on August 20, while making a run from Hog Island to Egg Rock, I saw a small, dark seabird on the water off Wreck Island, about three miles from Egg Rock. It turned out to be a puffin fledgling. It took off and passed our bow. I couldn't make out if it had a band, but I hoped it was one of ours, because it looked healthy and could already fly.

On August 23, another puffling scratched on the crew tent. Tom thought it had actually fledged the night before. On August 28, Tom scared yet another puffin out from under a rock. We thought it had fledged the night before, but it obviously lost its way, traveling a needless 150 feet through heavy raspberry vegetation.

But those were minor scares. Except for the transplant that died almost immediately after our trip and the chick that dug its way out of its burrow and was never found, all remaining ninety-eight of the hundred chicks matured to fledging age.

Two weeks after closing camp, I wrote Nettleship that we had fledged nearly 243 chicks since 1974 with a 98 percent survival rate. This was the final year of the original three-year allotment plan. I explained that although the time might have arrived to wait and see what we could learn from the transplants to date, I planned to have a team on the Egg Rock to watch for returns. Now that we had the technique and the new sod burrows, we could easily rear an additional age class.

7

Waiting

Months passed without communication from Nettleship. In February 1977, I sent him a follow-up letter and a proposal that asked for one hundred more puffin chicks to increase our chances of achieving the "critical mass" necessary for establishment of a new colony. But the reality was that we had no evidence that any of the chicks would ever come back. It was now four years since the project had begun and not even one sighting of a returning bird. We had nothing but hope to justify more puffin chicks.

When I finally heard from David, he apologized for his slow response and acknowledged our progress. But he seemed reluctant to commit to a transplant in 1977 and asked me to wait until May to discuss it. This threw me into a tailspin. How could I wait until May to plan for a summer program? I would need to line up funding, staff, equipment. And another year of permits would take time.

I was so committed to seeing the project continue that I decided to move ahead on the assumption that David would come through. I had no other choice. It was either move forward or give up and giving up was not an option. When we talked by phone in May, David at last agreed to another transplant if we would be willing to do some research on the response of birds to different kinds and amounts of food. He was interested in demonstrating the importance of capelin as a seabird food because there was a growing fishery focused on this key forage fish for seabirds and marine mammals. Our controlled growth data would offer some useful comparisons of capelin to other foods such as smelt and shrimp.

I agreed, and he approved the next hundred chicks. But clearly, the future of the project was not assured. If none of the 1974 or 1975 puffins were spotted in the coming summer of 1977, David seemed ready to declare the project a failure. The stakes were higher than ever for success: it was obvious that resighting some of our puffins in 1977 was now critical to the project.

I worried that these highly social birds might be returning to the vicinity of Egg Rock, but if they did not see other puffins on the island, they might not stay long enough for us to see them. And if they landed on the water but never came ashore, we would not see their bands and be able to prove that the translocations were working.

The urgency for success led to the genesis of our use of decoys. Waterfowl hunters have long used decoys to lure ducks and geese into hunting range, as did shorebird market hunters in the late nineteenth century. I also recalled a *National Geographic* article that showed a puffin hunter surrounded by dead puffins propped up as decoys. The hunter was using a long "fleyging" net to snatch puffins out of the air. The athletic-looking hunter, perched at the edge of the cliff with his long-handled net, would scoop up passing puffins much as a lacrosse player would snag a passing ball. This startling photo solidified my decision to use decoys for restoration. I reasoned that if decoys could lure birds into range for hunters, perhaps they could also lure birds back to historic nesting grounds.

I floated the idea by Kathy Blanchard, who was now working for the Quebec Labrador Foundation (QLF). She thought that the idea would interest Donal C. O'Brien, a talented decoy carver, whom she had met through her association with QLF. The year before, O'Brien had joined the National Audubon board. In his early forties, he was a respected Wall Street attorney whose extreme love of winter duck hunting along the dangerous, icy, and storm-lapped shores of Long Island Sound merited a 1971 *Sports Illustrated* feature story.

Part of the O'Brien reputation was his talent for carving award-winning duck decoys. He was a repeat US national amateur champion and notched two best-in-class honors at the 1971 world decoy championships. His favorites were "working decoys," a preference that predisposed him to designing functional decoys that would hold up in the field. He was enthusiastic about our decoy plan and happy to carve puffin decoys in two postures: standing and floating. With their flattened beak, puffins offered a special challenge, but Donal solved the problem by thickening the beak and neck to give the bird a stocky bowling pin look that proved successful in our harsh habitat.

To come up with the final prototypes, he collaborated with another Connecticut carver, Ken Gleason. The actual decoys that were deployed in the field were then replicated from the originals twenty-four at a time on a giant wood lathe. I thought it an interesting twist that this machine, which was developed originally for turning out rifle stocks, was now replicating puffin decoys that might help puffins reclaim one of their historic nesting islands. Kathy was now a PhD candidate at Cornell but

found time to help paint the decoys with several other volunteers in a warehouse at the Lab of Ornithology.

On June 3, 1977, Joe Johansen, assistant boatman John
Ryan, Kathy, and I headed out to open camp on Egg Rock with
forty-four floating design decoys aboard the *Lunda* and the
Audubon camp's *Osprey III.* At the island, we set about deploy-
ing the first twenty-four decoys of the floating design. These
were arranged four or five to a string and then tied to two
bricks that we hoped would hold them tight to the ocean floor.
The plan might have worked in a calm bay or a lake, but an
open ocean setting with ten-foot tides was very different. We
loaded the decoys into Joe's dory, and he set about rowing near
the island, while I dropped the strings of floating decoys over-
board into relatively shallow water around the south end of
the island. We watched the bricks sink to the bottom and were
pleased to see how convincing the decoys looked floating on
the surface. We had some doubts about the effectiveness of the
bricks over time, so we stenciled "Property of National Audubon Society, Medomak, Maine," on the bottom of each decoy.

All was going well until the wind quickly came up, interfering with the decoy deployment. Only about half of the floating decoys were in the water when Joe rowed with difficulty
back to the island to drop us off with camping gear and the
standing decoys. He then transferred the remaining floating
decoys that we had not yet deployed to the *Lunda,* which was
now pitching and tossing wildly on its mooring in the twenty-
five-mile-per-hour wind. With powerful rowing, Joe made it
back to the *Osprey* and took her back to the mainland, leaving
the *Lunda* on the island mooring.

Kathy and I focused on setting up the tent, which was now
very difficult, because the wind was gusting to thirty miles an

hour. After securing the tent, I walked back to the landing to check on the *Lunda* and was shocked to find that the boat had capsized and was now floating belly-up, still attached to the mooring! Some of our gear and a few of the puffin decoys were floating toward the shore. We hustled to salvage what we could, but most of the decoys and lots of other gear sank or floated off into waters unknown. I broke into someone's CB chatter and asked them to call Joe to explain our predicament. There was nothing he could do but call the coast guard, who said they would be out in the morning.

As promised, a coast guard boat arrived early the next morning, but their attempts to flip the *Lunda* upright with a grappling hook failed, only creating more damage to the boat before they were called off to an even greater emergency. Before leaving, they shouted that they would be back later. After seeing how little they could help, I was resigned to staying on the island and hoped that the next coast guard visit would prove more helpful. While waiting, I occupied myself by planting cherry tomatoes and removing several more clutches of gull eggs.

By late morning, another coast guard crew arrived in a more powerful boat. They tied the *Lunda* to the stern of their boat, revved up their motors, raced forward, and succeeded in flipping her right side up. They manually bailed her and then towed the crippled boat, with Kathy and me on board, back to Hog Island. We looked back at Egg Rock as we departed, noting that about three hundred black-backed gulls had already settled back on the boulders. When the *Lunda* was examined on the mainland, we found one floating puffin decoy in the hull, a small consolation prize.

When we returned to Egg Rock a week later, we found

two floating decoy sets tangled around lobster buoys. All of
the others were missing. We managed to untangle one set and
moved it farther offshore. The floating decoys looked amazingly realistic, but we concluded that they were impractical
and would likely require constant maintenance to keep them
untangled and clear of the lobster gear.

On shore, we mounted twelve standing puffin decoys on
high outcrops and boulders. We had designed a base consisting of a threaded metal rod attached to a four-inch-square
piece of plywood with a "t-nut." After threading the rod into
the t-nut, we slathered concrete mix over the base to hold it
to the granite. The black-backs must have done a double-take
of their own, because they started attacking the decoys, knocking them over with their feet before the cement could even
dry. But once the cement was dry, the gulls did not bother the
decoys any longer. In fact, they not only stopped their attacks
but also settled down near the models, completely ignoring
them. This moment of avian comedy was another example
of the intelligence of the gulls, which were quick to learn that
decoys are not worth their time of day.

This was a rough beginning, but the season was just starting, and we remained hopeful. On June 10, the misfortune continued when I tried to land Tom French and Dave Mehlman, a
new research assistant, on Egg Rock to begin their island stint.
Strong northwest winds made landing with the *Geezer* treacherous. Tom was able to row Dave ashore but he was half swamped
by the time he put Dave on the rocks. I attempted to maneuver
the *Lunda* closer to reduce Tom's rowing time, but a lobster
line wrapped around the *Lunda's* propeller. With her stern held
into the prevailing wind, she began taking on water. Most of
the gear was soaked, including sleeping bags. Tom managed to

dislodge the lobster trap, then rowed back to retrieve Dave, and we all went back to Hog Island to await calmer weather.

When we returned on the afternoon of June 12, we found placid water, a relief after our last aborted effort. Dave, Tom, and I approached our normal landing site on the north end of Eastern Egg and pulled up the mooring while Tom rowed Dave ashore. While Tom was rowing back to the *Lunda*, I spotted a quick-winged bird flying low over the water. Soon my curiosity was replaced by disbelief. It was a puffin! I shouted to Tom, who barely broke his stride to lift his ever-present binoculars. Tom didn't seem particularly moved, but I was beside myself with excitement. Later Tom explained that he and the others had just assumed that the project was going to work because I said it would, so they were not especially surprised.

We watched the puffin make several passes. After circling the landing area several times, it landed near the *Lunda* and started swimming toward us. It came to within ten feet of Tom, who was almost eye level with the bird while rowing the *Geezer,* and it approached within twenty feet of the *Lunda*. As it bobbed in the water, I studied it carefully with my binoculars. It had a small bill without a ridge on the orange part of its bill, which meant that it was just two years old. Older birds would have one or more bill ridges. That it was a young bird gave me hope that it was one of our puffins.

The puffin was amazingly unafraid of us, despite our screaming in disbelief and the churning of the motor. On the water's surface, it stood up and exercised its wings, then dipped its head in the water. Then it took off and flew around Egg Rock several more times before plunging into the ocean again.

A great black-backed gull noticed the lone, vulnerable puffin. Tom and I were aghast when it started diving repeatedly at the puffin, but each time the puffin slipped under the water to safety. The frustrated gull at last called off the assault, and I edged the *Lunda* forward for a closer view. Surprisingly, the puffin didn't dive but watched me curiously. I wondered if it recognized me for engineering its first trip to Maine, but I was more curious about its legs—was there a band? As I edged closer with the boat, I could not believe how close the bird was permitting me to approach. Finally I was close enough to see its left leg under the water—it was banded! It seemed to be a light green band, but the water could have easily distorted a white band to look green. I slowly backed off and buzzed back to the mainland to share the news.

I zoomed back to Hog Island in record time, tied up to the mainland float, and raced up the hill to find Kathy, who was preparing to leave for another summer with the Quebec Labrador Foundation. Joe was mowing the lawn in front of the warden's residence, but I shouted to him in excitement over the roar of his mower.

"A puffin has returned!"

Kathy heard the hubbub and dashed downstairs.

"Quick! Let's go see the puffin," I shouted. I grabbed my four-hundred-mm lens and camera, and we raced back to Egg Rock. It had been more than an hour, but the puffin was still where I had left it. It was playing with a lobster buoy, picking curiously at the bobbing handle and frayed rope end, oblivious to what its presence meant to me. I approached for several photos and then backed away. I hoped it would see the decoys and decide to come ashore. Kathy and I headed back to Hog Island, leaving Tom and Dave on Egg Rock to keep an eye on things.

The next morning, Tom and Dave reported that a puffin was sighted to the northwest of Egg Rock. They noticed it because a lobsterman was already pointing at it. We continued mist netting and banding storm-petrels, warblers, and swallows and hunted for more storm-petrel burrows. One night among the 102 storm-petrels netted, 78 were previously unbanded. One veteran petrel was particularly special—it was the very first banded in our occupation of Egg Rock, the one that had flown into Kathy's tent three years before. Such discoveries offered us great delight.

Then, ten days after our lone puffin was sighted, a lobsterman pulled up to Eastern Egg Rock shortly before noon to tell us he saw a puffin that morning off nearby Western Egg Rock. As he spoke to us from his boat, a puffin showed up and circled him. It made appearances on all sides of the island throughout the day and flew close enough that one could see the white band on its left leg—proof at last that this was a two-year-old translocated Newfoundland puffin! While bobbing on the water, it nibbled and played with lobster buoys. On several occasions the naive puffin was seen floating within two or three feet of great black-backed gulls, but happily, the gulls were uninterested.

The next morning the puffin was still around, picking at colorful lobster buoys. It generally stayed within a hundred feet of shore before eventually flying off. Inspired by the sighting, we put up ten more "standing" decoys on the boulders. A week later, Dave was making pancakes for breakfast when he saw two puffins fly by the campsite. Eventually one of them landed among the decoys on Gull Rock! This was our first puffin on land and cause for great celebration. Then the second bird plopped down near the first. Dave and Tom detailed every behavior, including how the puffins cautiously approached the decoys, bowed, picked at decoy beaks, and nibbled at their wooden bellies. We could only imagine what they were think-

Puffin decoys encourage returning puffins to come ashore, land, and become more familiar with new nesting islands. Once ashore, the gregarious puffins help to attract others. Photo by Derrick Z. Jackson

ing about this strange island. But the main thing for now was that they had come ashore.

Tom was the first to confirm that one of the birds had a white band on the left leg and a metal band on the right leg. Surprisingly, the other was unbanded. He and Dave tried to get close enough to take photos of the band, but the birds became nervous, so they retreated back to the tent and pulled out their spotting scope, hoping to read band numbers. The pair flew away when a black-back cruised low over their heads.

Because these were the first puffins to land on Egg Rock, every detail about their appearance and behavior seemed enormously important. To our knowledge, no one had ever watched the genesis of a puffin colony before. Not knowing what was really important and after waiting four years for these encounters, we recorded every detail. We noted that their bills were roughly 50 percent larger than those of the decoys and that real puffins leaned over considerably more than the (completely erect) decoys. This made the live birds shorter than the decoys, a useful fact for us in picking out the live puffins at a distance. The banded puffin led the other bird in "touring" the island, and on several occasions the unbanded bird remained on Gull Rock while the banded one circled around alone. They noted a fleeting glimpse of billing behavior, which suggested (we hoped) that the birds were already a pair. Our optimism soared with these first puffin flights!

We reflected that this was the first recorded landing on Egg Rock of an Atlantic puffin since the 1880s, and I was beside myself with joy. Although we had waited four years for this moment, the thought of puffins coming back on land, and then edging into nesting habitat, was exceeding my best expectations. In heavy morning fog on July 2, Dave spotted a puffin flying over the west shore, circling Egg Rock many times before it settled down on the water and drifted in the current. The Hog Island campers were coming out to visit, so everyone kept their fingers crossed that the puffin would hang around.

It would be the first time we could "show off" a puffin as evidence that our project was having success.

The puffin cooperated, much to the delight of the campers, who broke out in spontaneous applause when it came into view. This was a huge milestone for me, as one of my initial thoughts about starting Project Puffin was to give the Hog Island campers a chance to experience this charismatic seabird. The lone puffin, as if starved for social interaction, bobbed its head at an orange lobster buoy before it flew off to the south.

Four days later another lobsterman blew his horn at Egg Rock. He was holding up one of our floating puffin decoys. A string of five had broken free in the turbulent waters and he happened upon it in the course of pulling up his traps. The lobsterman threw it toward us as if it were a football, but it fell far short. Finally, Dave snagged it with a large fishing lure.

In the coming weeks, we had other messages from floating decoy finders. Nearly all of the "floaters" were now at large in the North Atlantic. A few were recovered washed up on beaches in southern Maine, and one was found in Provincetown, Massachusetts, where it was held hostage as a doorstop, but we were happy to know it was in a good home and making friends for puffins.

Another returning puffin was spotted circling Egg Rock on July 8, but Tom, Bruce, and I had to step away from the exciting events at Egg Rock to collect the next hundred chicks.

The weather in Maine was crystal clear for takeoff, but as we neared eastern Newfoundland and the Saint John's airport, we entered heavy fog. The airport control tower began talking to the pilot of our small prop plane. Every few seconds we heard "a little bit to the left," then, "a little bit to the right," then, "a bit up," then "a bit down."

Suddenly, we heard, "Pull up! Pull up! Pull up!"

When I looked down, I could see the details of building roofs—shockingly close.

Our pilot quickly gained altitude and asked to try again. The tower said "NO!" and sent us north to the Gander airport, where we had to rent a car and drive more than two hundred miles back to East Bauline. Fortunately, we arrived in time to meet John Reddick, who took us out in the late afternoon. We went to sleep on Great Island to the calling of thousands of storm-petrels. The next morning, the puffin chick collection went so smoothly that we had a few hours to photograph seabirds.

The flight home was mercifully calm, and we placed another set of puffin chicks in our sod puffin condos by midnight. The sod puffin burrows were proving to be a success beyond our highest expectations. They were cool even on hot days, and the chicks were growing up clean in burrows that they could customize by digging. We continued to improve our food supplies and obtained frozen silversides that were large enough to accommodate vitamins and just the right size for chicks to swallow.

The only bad news was that in the first couple of days one chick died, apparently choking on a large silverside. The sadness of the death was lightened a couple days later when our mist nets captured our first Baltimore oriole in four summers. That we still had time to mist net and band land birds at Egg Rock in these years speaks to how few seabirds were present on the island. Banding helped to demonstrate the broad use of the island by land birds (a total of 215 bird species—mostly migratory land birds—have been tallied from this seven-acre island).

Having time for projects such as banding land birds also

Tom Fleischner prepares a batch of thawed silversides by counting out meals and inserting vitamins in the mouths of the fish (inset). His worktable was built of the repurposed ceramic burrows from the previous year. Photos by Stephen W. Kress

meant that we had so far been more successful depleting the gull population than restoring any of the lost species that we hoped to attract. Most of the nesting gulls had been chased off by 1977, and days would pass without a puffin sighting. Feeding puffin chicks twice a day took only a few hours; we thus had time to "ring and fling" a few land birds and storm-petrels at night.

We would also distract ourselves with other wonders of nature. One day, Egg Rock was surrounded by three minke and eight pilot whales. Another day, thunderstorms and lightning with the rhythm of fireworks gave way to a rainbow stretching from one end of the horizon to the other, with the ends of the rainbow descending over more whales. On still others we appreciated nature's bounty for our dinners. Without many puffins to watch, we supplemented our diet of peanut butter and freeze-dried food with cod and pollock.

Although the puffin sightings were few, the presence of the bands on the puffins bolstered our conviction that we were well on our way to restoring a puffin colony. No one else in North America was putting white bands on puffins, so we knew that these were "our" puffins and that we were witnessing the earliest events leading toward the rebirth of a puffin colony. There was also the practical need to capture as many of the sightings as possible to demonstrate to Nettleship and our supporters that the project was working.

It made sense that our first returning puffins would be wearing white bands, even though we had hoped to see one of the five green-banded puffins from 1973 or one of the fifty-four blue-banded tile-reared puffins from 1974. But the chances for resighting these were slim at best. The 1973 birds had spent most of their prefledging days on Hog Island, and there were just five: the odds were against them. Likewise, we suspected that the plumage of the 1974 puffins was compromised by the tile burrows; and they certainly experienced more stress than birds moved in later years. In contrast, there were ninety-one chicks fledged in 1975, and most of these fledged in good condition from soil burrows, despite the flooding trauma of that summer.

The puffin resightings continued, each one logged in detail. Tom also noted what would become a common birding confusion at Egg Rock. He described a sailboat that circled the island while one of the puffins was floating off the south end. A woman onboard was scanning with binoculars. She overlooked the living puffin but caught sight of a lone floating decoy, grabbed a camera with a telephoto, and snapped a picture. Tom observed that the boaters seemed even more excited when they passed Gull Rock and saw the whole flock of standing decoys yet missed a live puffin that was sitting in the water nearby.

The next day, I brought Carl and Harriet Buchheister to Egg Rock to show them the array of decoys and to spend a night on the island. I also brought out Bill Bridgeland, a lanky naturalist friend from Ithaca, who was now helping us on the island. While still onboard the *Lunda* and bundled from waist to head in a thick blanket (her usual boating attire), Harriet managed to be the first to spot a puffin. With glee, she shouted, "There goes one!" Carl immediately followed, "It looks dark in the face," which would indicate a young or even just one-year-old puffin. The puffin landed in the water a hundred yards off the southeast shore, exercised its wings, then rose up again and flew past us as if to provide a better view of its gray face and dark bill with just a bit of dull orange at its base.

Having the Buchheisters visit was a great pleasure, but it forced a sense of formality that was largely foreign to this simple outpost. It was clear that they were not comfortable with the co-ed living situation, but they kept their focus on the project and birds. Although most of us were used to sleeping in the same tent and using the same open-air privy that Joe Johansen nicknamed the "Thunder Mug," seventy-something Harriet was very old school. And she was anything but shy about her needs. At her request, we set up a separate tent for Carl and her. We also added a temporary roof for the privy, because the few nearby shrubs provided less than the desired level of privacy.

Egg Rock camping reminded Harriet of a previous over-night she spent on a seabird island. Years earlier Helen Cruick-shank and she went to Funk Island in Newfoundland with Roger Tory Peterson. The island is famous as the last North American nesting place for great auks but is now a vast murre, gannet, and puffin colony. The women brought their tent but Peterson did not, thinking he would gaze at the stars from his sleeping bag. That night it rained in torrents. Peterson came to their tent to seek shelter, and at first they would not let him in. They finally took a measure of mercy and said he could come in halfway, down to his waist. His lower half had to en-dure the rain.

Now, Harriet offered to cook dinner for us. At 5:20 p.m., as she sautéed zucchini, Bill saw a puffin zip right across the island. It landed in the water between the shore and four float-ing puffin decoys (the last surviving string). The puffin was so tame that Bill was able to stalk within thirty to forty feet of the bird—so close that he could easily see a white band on its left leg.

After dinner, at about 7:40 p.m., Tom and I walked up to

Flagpole Hill, where we spotted the puffin standing next to the lower decoy on Gull Rock. Tom, Bruce, and I all took pictures in the fading light. As it got darker we watched the puffin walk up and spend several minutes in the middle of the large group of decoys on top of Gull Rock. We banded fifty of the transplanted puffin chicks on August 8, and most looked as if they would fledge in just a few days.

The same day the seas around us churned with nearby drama. Tom and Bruce spotted a large shark and then a second cruising just off the west shore. Using lobster buoys, gulls, and cormorants for comparison, Tom thought that the sharks seemed to be about five to six feet long. They cruised slowly on the surface with their dorsal and tail fins showing, the tail fin constantly pushing from side to side. One shark entered a school of pollock and began to thrash the surface, even lifting its head out of the water.

At that moment three black-backs appeared; one dropped right in front of the shark and grabbed a pollock, then took off. Such shark sightings were rare near Egg Rock, but it reminded us that the puffin fledglings might also become shark meals, especially if our shiny metal bands became fishing lures. There was little we could do about most of the off-island threats.

August 8 saw the banding of the last of the surviving ninety-nine puffins. They all looked good to go, and departures indeed began before dawn on the ninth. As Bruce and Tom inspected burrows to see which chicks fledged, a lean, rufous-colored bird flushed out of the bushes and plunged headfirst into the mist nets. It was a least bittern, a secretive member of the heron family that hunts in the reeds. This was another first for us, yet another bird to find oasis during migration.

On August 9, I brought Bill back for his next island stint

and took Tom back to Hog Island. Soon after we departed, Bill spotted a large brown bird flying right behind Bruce at about waist level. Bruce wheeled and yelled, "An owl!" Just then the bird turned toward Bill, who froze as it flew straight at him. With nothing short of pure predator instinct, Bill leaped on the flying owl and the two tumbled into the nettles. Bill emerged holding the owl, but the owl was holding him too, having locked onto his hand with its long talons. With the owl hissing and snapping its beak, Bill pried the talons out of his bleeding hand and examined his catch, a very perturbed long-eared owl. We could only wonder what this island had in store for us next.

When they called me on Hog Island, I asked them to band it and hold it until I arrived the next morning. Tom joined me for the trip back to Egg Rock. While Tom was handling the owl, a talon punctured one of his fingers. Bill was still nursing his wound from sacking the owl. Both of them ended up making hospital visits for severe staphylococcus infections. They were both placed on an antibiotic IV; Tom almost lost part of a finger.

Meanwhile, the puffin departures continued, with about a fourth of the group fledging the night of August 14. As I took a recovering Bill (still taking penicillin four times a day) and Bruce back to Egg Rock, we spotted one of our new fledglings with this year's yellow color band floating in the water, swimming with a strength that belied its youth.

The nighttime puffin parade astonished famed British ecologist Mary Gillham, who was our special guest for three days. In her guest journal entry, Gillham was fascinated with our sod burrows. She noted how they were "painstakingly made by hand—a most enterprising project." She told us that

in Wales, puffins and other auks were declining due to an explosive increase in gulls—much as in the United States. She
noted that there was much to learn from our "puffin rehabilitation study. Perhaps we could emulate it."

An incurable outdoorswoman (one of the first women to
participate in an Antarctic expedition in 1959), Gillham praised
the simple Egg Rock lifestyle: "Some folk come all the way from
Britain just to savor the sites of New York and Washington—
those poor misguided souls. This is the real America."

Two weeks after Mary's visit all of the chicks had fledged,
and I wrote David Nettleship that we had just completed another successful summer rearing puffins on Egg Rock. Ninety-
nine of the hundred chicks made it off the rock in good order.
I explained that even more exciting for us was the return of
some of the transplanted birds. It was difficult to know how
many individuals the twenty sightings represented, as all of the
birds wore white bands, indicating they were from the 1975
group. We would have needed to read the metal USFWS bands
to recognize how many individuals the sightings represented,
but we did not want to approach that close. I explained that we
also noted visits from at least one unbanded puffin and a dark-
beaked bird that may have been a one-year-old.

The white-banded puffins eventually lost interest and in
later years were not known to return to the island. Instead, they
gravitated to established puffin colonies at Matinicus Rock
and Machias Seal Island. One even nested on Pearl Island,
halfway up the Atlantic coast of Nova Scotia. But the returning white-banded birds in 1977 gave me hope that the project
would eventually result in a new colony, and they also convinced David Nettleship to continue helping by agreeing to
provide an additional allotment of one hundred chicks for each
of the next four years.

Our excitement about returning adult and young puffins made me set up camp on Egg Rock much earlier in the spring of 1978—on April 23. I didn't want to miss even a single sighting.

Even though we had no proof that puffins would actually nest at the island, I was beginning to envision a natural conclusion of the project. My hope was that puffins would nest in rock crevices, and our manmade sod burrows would eventually collapse back to the earth. The tent site would come down and the people would leave. But there was one problem: the herring and great black-backed gulls.

I wondered how the puffins could survive here without our chasing off the gulls every year. I considered the puffin and tern colonies at Matinicus Rock and Machias Seal, because I always thought of these islands as my model for sustainability. I revisited Palmer's history of Maine birds and, sure enough, found accounts that Egg Rock was once home to a large colony of common and Arctic terns; some still nested here as late as 1936 when the Audubon camp opened. Egg Rock must have been like a little Machias Seal Island—complete with puffins AND terns. I reasoned that if a large colony of terns could be restored, perhaps they would drive off the gulls and the puffins would live under this protective umbrella, eventually permitting us to pack up and leave the island to the birds.

I asked Donal O'Brien to turn his carving talents to Arctic tern decoys. He promptly provided two elegant poses, an incubating and an alert posture, and these were replicated, twenty-four of each design. I also decided to play tern colony sounds as a background ambience. I was very impressed by the attraction power of storm-petrel recordings and wanted to see if tern recordings would have the same effect. Tern colonies are noisy places, and I wanted to replicate the sound of a thriving colony. I reasoned that a quiet tern colony is generally a dangerous place, because terns go silent just before the whole colony flies up on the approach of a predator. On this hunch, I chose tern courtship sounds rather than the aggressive sounds that one hears when walking into a tern colony. We had a ready source for such recordings at Matinicus Rock, and I made my first Arctic tern recordings from observation blinds to minimize the aggressive calls that parent terns give when people walk near nests.

Tern restoration had its roots in attempting to build sustainability for puffins, but it soon became apparent that tern restoration was needed for its own sake, because Maine tern numbers had been in a slump since reaching their twentieth-century peak in the 1930s. But this idea was easier in concept than reality.

Technology had not quite caught up to my plan. Before the advent of MP3s and DVDs, the only remote playback technology was cassette tapes. Endless-loop cassette tapes were commonly used in home telephone answering machines at the time and proved the obvious solution. I found that I could make a recording and play it through a car cassette player already designed for playing off of twelve-volt DC power. Most of the necessary parts were available at the nearest Radio Shack. This was all good technology for my plan, but it meant moving a fully charged car battery into the tern habitat, and there was no easy way to do this, considering the difficult landings and long treks overland to the best habitat. Despite this obstacle, for many years we managed to drag the beastly heavy sixty-pound batteries ashore with the *Geezer* and then moved them over slippery intertidal rocks and across the island.

Given the difficulty of moving batteries, we decided to ration the power. In early June 1978, we set out twenty-eight realistic wooden tern decoys in a gravelly area on the south end of Eastern Egg Rock and installed a sound system that projected nonaggressive Arctic tern calls. To save power, we turned it on only when we chanced on a tern near the decoys. We noticed, however, that terns were more likely to stay when the system kept running. Photovoltaic solar panels (the obvious solution) were still several years away.

The initial signs were promising. Soon after the decoys were in place, fifteen terns settled into the area and some hovered over the loudspeaker. With their streamlined shape, they are built for the sky, where they spend much of the year on the wing. At the time, it was only guessed from a scattering of band returns and winter sightings that the Arctic terns that nest in Maine were flinging themselves into the global wind stream in the fall and traveling to Antarctica for the winter. Later studies at Egg Rock would show that the actual round-trip flight paths from Maine to Antarctica and back to Maine could take the birds via Africa and Argentina on trips that could exceed forty-four thousand miles per year. We were thrilled by the increasing number of terns we saw, especially when males appeared with fish to offer to potential mates, although it was a little ridiculous when the males offered the fish to decoys and attempted copulations with the models.

But arriving in April meant that we had to suffer the worst of Maine's long, cold spring. Much of April and May was spent bailing out the tent and recovering from fearsome storms that could move boulders weighing several tons. Joe Van Os, a birdlife instructor from Hog Island (later to become the owner of the world-class Joseph Van Os Photo Safaris), and Lin Peyton, a camper with an interest in the ministry, agreed to volunteer during early spring. On May 9, Joe and Lin endured an especially challenging day.

In the daily journal, Joe described how the day started with a steady rain at night and then "aired up" about daybreak and did not let up until 6:00 p.m. With the drenching rain and winds gusting to forty-five miles an hour, the leaking twelve-by-twelve-foot wall tent collapsed on the intrepid team, already standing in several inches of water that had collected on the tent floor. Determined to salvage what they could, they bagged everything possible into plastic bags—including their shotgun, books, and records. They didn't bother to save the bags of cement mix intended to mount decoys: these were already soaked and ruined by the deluge. They extracted themselves from the collapsed tent and made a makeshift rope rig to pull the tent back up, but in the process their Coleman lantern (the only source of heat) fell and smashed on the floor. Their sleeping bags were soaked, and they were generally cold and semi-miserable.

Their journal notes described that as the rain and wind continued, waves broke over the center of the island and swept at least two decoys off Gull Rock, one of the highest points on the island. On seeing this, they went out to retrieve the remaining puffin decoys and piled up the tern decoys. Miniature lakes about eight inches deep now sprang up all over the island. Standing on Flagpole Hill was difficult due to the wind, but from this higher vantage, they could see birds that typically frequented the rocky shoreline, such as semipalmated sandpipers, semipalmated plovers, and purple sandpipers, seeking shelter in the flooded raspberries!

On returning to the tent, they found the CB radio wet, and when they turned it on, a large puff of smoke came out of the back, giving Joe a startling shock. Yet they were able to radio the mainland and talk to Mary Johansen to say that they were okay, but not to expect another call until things dried out. Three days later, Lin commented in the journal: "Cleaned out tent. Ready for next crew. One fabulous stay."

The stormy spring continued. On May 16, the rock was blasted by a nor'easter so strong that Tom, Kathy (who had returned to help us start the new season), and new researcher Richard "Rich" Podolsky reported that they began feeding bread crumbs to shivering sparrows outside the tent, bailed twenty-five gallons of water from the inside, and salvaged paper records by stuffing them inside trash bags. The scene got gloomier as food, cots, camera gear, and papers began soaking up the water.

The all-too-familiar plan when a storm hit was to bag everything in sight and to bail buckets of water from the flooded tent. This was quite the introduction to island life for Rich. At this point we had the luxury of a small Coleman cooking stove but used it more to take some of the chill from the air. The tent interior was quite a site—three researchers huddled together in damp sleeping bags like so many sardines—surrounded by dozens of green plastic garbage bags. Together, they waited for the end of the wet onslaught that lasted through the night. Their companions for the night were a few soaked song sparrows that were given refuge within a bucket near their bedside.

I returned to Egg Rock on May 22, with new research assistant Evie Weinstein. Evie and I met in Ithaca the previous winter and were beginning a relationship. I was eager to share the Egg Rock experience with her and arranged for a few days on the island, relieving Tom, Kathy, and Richard, who departed for the mainland. They had left the camp in decent shape, con-

sidering the bedlam caused by the recent weather. We were
fortunate to have relatively mild weather for our stay.

Some days later, two volunteers, Susanna Davy and Marge
Winski, took up caretaking Egg Rock, and I returned to Hog
Island, where I would check in a couple of times per day. From
my desk in the Binnacle, I could usually talk to them by CB
radio. On their first radio check, the reception was especially
bad and crackly. As was often the case, I could hear chatter
from Puerto Rico better than from Susanna and Marge just
eight miles away. As they gave me their latest observations of
terns, my pulse quickened. In the garble I heard something
about eggs. I shouted, "You have four terns laying eggs?" As
was often the case, the CB radio left much to one's imagination. Perhaps it was her thick English accent, but Susanna was
actually trying to tell me that there were two terns sitting on a
rock at the north end of the island and one was giving a courtship display. Even this was still big news, and I was eager to
hear every detail.

I was beginning to doubt the value of the early startup,
considering the extreme weather and the fact that the teams
had not reported a single puffin sighting. But that changed on
May 28. That evening a puffin appeared, then flew off far onto
the water. On June 3, two puffins arrived together. From then
on, sightings became regular. Then something happened we
had not seen before. At 6:00 a.m. on June 8, Susanna and
Marge saw a puffin with a white band engage another with a
courtship display.

They called me on the radio at 7:30 a.m.: "We are so excited we can hardly hold the microphone." When they told
me what they had seen, I asked them if they were having hal-
lucinations from having eaten Marge's seaweed pudding—a

favorite island treat. We laughed, but we all knew that we were seeing a miracle unfold.

By 1978, it was now the fifth year since we had moved the first puffin chicks to Eastern Egg Rock and still no nesting. Terns were making only the most fleeting visits. As persistent as the gulls had been, our concerted effort had broken their hold on the island—they were gone—but we had not yet convinced either the terns or puffins to commit to a homecoming. Five years into the project, and I sometimes wondered if our efforts would only confirm that people could easily deplete wildlife. Reinserting missing species was clearly a much more difficult process.

Hope returned on June 19 when Saskia Franzeen, a new assistant, wrote, "Terns galore on Eastern Egg Rock today! At 11:40 we counted about 30 fishing along with 8 great black-back gulls and 10 double-crested cormorants." Late in June, another new research assistant, Tom Fleischner, said the terns edged closer to mating as a pair in the decoy area fed each other. He noted that it was another good sign that the terns were becoming increasingly aggressive, defending more and more of the island. Tom was thrilled to report that he was chased from both Flagpole Hill (on the south end of Egg Rock) and from the northeast corner of the island in the same day. He also reported seeing a tern chase a black-back away. These observations were big news in our little world.

Despite the occasional sighting of terns copulating, we were in an eight-day drought of puffin sightings. When I finally saw one on the ninth day (during a brief supply delivery), Tom mused that the spotting raised "silent questions in our minds" whether the newer, less experienced team had been missing puffins all along or whether there were troubling unknowns that had them missing in action. Sightings remained scattered right up to the time to fetch another hundred chicks. On July 8, I returned to Egg Rock with Richard and Susanna Davey to assist Tom with final preparations for the 1978 transplant. We spent most of the night doing some serious storm-petrel netting, so we were sleeping late when Susanna came running back to the tent. Her British locution could not be restrained. She shouted: "Steve, *Lunda*—she is sinking!" Susanna was up early doing the daily 6:00 a.m. bird count when her binocular scan fell onto the *Lunda*. The bow and little else was above water.

Tom, Richard, and I raced to the landing. Sure enough, most of the *Lunda,* including her 90-horsepower motor, was underwater, and it looked as though the remainder was about to go down. We launched the *Geezer* and bailed water for forty-five minutes. The large outboard was soaked and would not start, but the 7.5-horsepower backup reluctantly started.

Tom, Richard, and I spent most of the rest of the day inching our way back to Hog Island, happy to have not lost the boat. Later, we surmised that someone had left the drain plug out while the boat was moored.

Rich, Tom Fleischner, National Geographic photographer Jonathan Blair, and I flew to Newfoundland on July 11. By 2:00 a.m. of the thirteenth, twenty-three hours after starting the collecting, we had tucked the puffin class of 1978 into their new burrows at Egg Rock. The day after a transplant we all glowed with a sense of accomplishment and exhaustion—knowing that we had achieved something very concrete and full of hope by bringing another puffin class to Egg Rock. Tom noted, "Most of all I have great respect for these little critters that can take so much craziness so well—probably better than we could."

Like the previous year, we had an early loss. The chick in burrow 31 was missing, with a pile of uneaten fish left behind.

There was a hole in the rear of the burrow. Tom and Rich beat their way through the raspberries on hands and knees searching for the chick to no avail.

Then Tom found a lifeless chick in the back of burrow 67. These losses and several days of rain lowered spirits. Everything was damp and gray; Rich and Tom blamed themselves for the deaths.

They discovered a hole between burrows 43 and 44 and found another dead chick. It was the one that had been in burrow 44; it had gone through the hole only to starve while the resident puffin chick ate all the food.

While I encouraged Rich and Tom to focus on the ninety-seven chicks that were thriving, I shared their frustration that three chicks had died. We found some solace, because, despite these losses, the translocated chicks had a much better chance of making it to fledging under our watchful care then those back at Great Island, where only about 37 percent of pairs fledged a chick—largely because of predation from herring gulls.[1]

But the bad news continued the next day with the discovery of two more dead chicks. At first Rich and Saskia thought they had choked to death; then Rich speculated that the fish might be too large for the smallest chicks to swallow. The next day, a bird disappeared out of burrow 38. Then yet another dead chick was found in burrow 71.

That meant a total of seven dead or lost chicks, a record loss for us. In the three previous years combined, we had lost only five of the nearly three hundred transplanted chicks. Despite the losses and low spirits, the team kept busy doing whatever we could to help the remaining birds thrive and keep a positive attitude. Rich clipped a pathway from the burrows through the brush and down to the ocean, which he hoped the

Every effort was made to minimize contact with translocated puffin chicks. Tom Fleischner and other interns quietly placed puffin chick meals in the entranceway to the burrows and tallied discarded meals. Photo by Stephen W. Kress

chicks would use on fledging night. I continued to bring Hog Island campers onto Egg Rock to show off the healthy chicks.

On August 1, an adult puffin showed up on Gull Rock. It spent five and a half hours on the boulders or in the water just down below, long enough for us to get a look at its band. It had a black leg band, the color we used in 1976. This was the first confirmed returnee from that year. Moreover, it wandered close to the burrows with our Newfoundland chicks. Did it sense the chicks inside of them? Or did it remember its own sod home from two years ago?

Tom spotted two more puffins through his spotting scope, one of them a dark-faced immature that was in the company of a 1975 transplant, recognized by its white band. We

After Marlin Perkins expressed disdain for riding in the *Geezer* and recommended rubber landing boats, we switched to inflatables—and named the first such vessel the *Minke* for the gray-backed whales that sometimes show themselves near Eastern Egg Rock. Without a dock, nearly everything lands on Egg Rock via the *Minke*. Here Juliet Lamb rows the *Minke*. Photo by Derrick Z. Jackson

could not see if the younger bird was banded, but we were surprised to have such a young bird already back at the island. But then we were becoming accustomed to surprises and happy to have a puffin of any age showing interest. After nearly two weeks without a puffin, this was the first time we had three different puffins on the island on one day, and at least two that were ours from the transplants.

The banding of the "Class of '78" began in earnest around August 10. We used a black-and-orange-striped plastic band (which we called bicolor) provided by British puffin expert Mike Harris in combination with five-mm-high solid color

Male and female puffins look similar, except that the male of the pair (right) is usually a little larger than the female. Pairs are usually monogamous, returning to the same burrow for a decade or more. Photo by Derrick Z. Jackson

These fish will soon become food for a hungry chick. The pink-bodied fish with the forked tail is a haddock; the smaller fish with the pointed tail is a redfish (*Sebastes* spp.). Haddock were unknown in the puffin chick diet before 2010, when they suddenly made up about 25% of the fish delivered to Matinicus Rock chicks. Redfish (ocean perch) were first observed at Matinicus Rock in 2011, comprising 25% of the fish delivered to chicks. Both fish are now chick diet staples. This development is notable because both species, previously overfished, are now recovering, in part due to effective fishing regulations imposed by the Magnuson-Stevens Fishery Conservation and Management Act. Photo by Derrick Z. Jackson

In 1977, puffin decoys were set out at Eastern Egg Rock to encourage puffins to land on the island; today, decoys encourage puffins to frequent locations where observers can read leg bands and note the kinds of fish (such as these herring) that are fed to chicks. Photo by Derrick Z. Jackson

Puffins typically sit in tight groups on nesting islands, a behavior that provides safety from such predators as great black-backed gulls and peregrine falcons. If a predator were to attack such a group, the puffin flock would scatter, confusing the hunter. Photo by Derrick Z. Jackson

The puffin's upright posture, round shape, and bright colors give it a compelling, clownlike appearance, but this is all coincidence. Puffins stand upright because they use their feet as rudders when swimming. Likewise, their round-bodied look results from massive breast muscles that power their flipperlike wings underwater. The bright colors indicate age and breeding condition. Photo by Derrick Z. Jackson

Although puffins prefer the safety of numbers when sitting on land, they also keep personal distance from others—usually a respectful pecking distance from their nearest neighbor. Photo by Derrick Z. Jackson

Puffins are famed for bringing multiple fish to their chicks at once. This permits them to forage miles away from their nesting island and to efficiently bring many fish at once rather than making multiple trips. To accumulate their impressive catch, they hold the first fish tight against the roof of their mouth with their tongue while deftly dropping the lower mandible to snatch additional fish. Photo by Derrick Z. Jackson

Northern gannets once nested in the Gulf of Maine, but nineteenth-century hunters wiped out these southernmost colonies, which have yet to repopulate former nesting places. Yet gannets are still common birds during summer months in the Gulf of Maine, and they sometimes come ashore on puffin nesting islands. Photo by Derrick Z. Jackson

Common terns, recognized by their black cap and black-tipped orange bill, nest on all Maine puffin nesting islands, where they help to drive off predators such as gulls and raptors. White hake, a coldwater fish whose present range is shifting northward as a result of climate change, is presently the most common food fed to tern and puffin chicks. Photo by Derrick Z. Jackson

Black guillemots are perky relatives and neighbors of puffins. Like puffins, they nest in dark rock crevices on remote islands. This guillemot has captured a black-bellied rosefish and paused for a moment before scrambling into a burrow to feed its chick. Guillemot chicks served as surrogates for puffins in 1972 and 1973 to help develop rearing techniques for puffins. Photo by Derrick Z. Jackson

Atlantic puffins acquire their famed rainbow-colored bill when they are two years old. At three or more years they obtain raised ridges on the orange part of the bill; this puffin is at least four years old because it shows two such ridges. During winter months, the courtship colors of the bill and eye and orange patch of skin at the base of the beak (rictal rosette) are lost. Photo by Derrick Z. Jackson

coils, putting as many as four bands on each chick. We hoped these would help us recognize individual puffins by their color combinations. Previously, entire year groups were banded with the same color, and we were frustrated by not being able to recognize individuals at a distance. We called the class of 1978 simply "the bicolors." This seemed to be a good system, but we discovered years later that with four bands, the odds of losing one or more were much greater than with a single band and led to much confusion. Trial and error continued to be our operating system.

8

Triumph or Tragedy?

Although the occasional visits of puffins kept us hopeful about the chances for eventual nesting, there was growing concern that puffins were not finding Egg Rock a suitable site. Six years into the project, I was beginning to lose sleep wondering if our Newfoundland hatched puffins would just tease us with visits to Egg Rock but eventually settle into the colony at Matinicus Rock with the other puffins rather than pioneer a new colony. I considered not asking Nettleship for another hundred chicks from Great Island. I knew that at some point we were going to need to step back and just watch for the outcome.

Then Don Meier, producer of Mutual of Omaha's *Wild Kingdom,* approached me. At the time, this was the premier wildlife documentary series. Marlin Perkins, former director of the Lincoln Park and Saint Louis Zoos, had built a reputation with sidekick Jim Fowler of capturing wild animals for display in zoos. But by the late 1970s, subduing animals with nets and brawn was falling out of favor and *Wild Kingdom* had shifted to documenting conservation stories. The *Wild Kingdom* team had just completed a film about Cornell's peregrine restoration project, and the peregrine team had suggested that they do a film about puffins.

This led Meier to approach me about producing a sixty-minute television show about Project Puffin that documented the collection of puffin chicks in Newfoundland as well as the chick rearing at Egg Rock; they were willing to cover all of the expenses for the 1979 transplant. The trip was arranged with David Nettleship, who generously agreed to have Marlin and his film crew join the 1979 Newfoundland puffin collecting trip.

As usual, there was much to do at Egg Rock to prepare for the coming season.

I thought it was time for some improved housing for the Egg Rock team. It was clear that tents were no match for the extreme weather. I reasoned that solid walls would break the relentless wind and boost morale. I was also coming to realize that, if we were eventually successful in starting a puffin and tern colony, we would need a long-term presence on the island to guard the colony from predators. This was a shift from my earlier vision that we could bring puffins back and eventually withdraw all signs of humanity.

We had permission from the owner of the island, Maine Department of Inland Fisheries and Wildlife, to build a temporary shelter for the island team, which I interpreted as a twelve-by-twelve-foot cabin, built with bolts so that it could in theory be disassembled at some future time. Recalling the extreme weather encountered in April of the previous year, we decided to push back the startup of the 1979 season to late May. Joe Johansen designed and prefabricated the Egg Rock cabin on the mainland over the winter months. On May 22, we were ready to start the 1979 season. I would soon be reminded that summer in Maine doesn't really begin until July.

Joe gave us a wake-up shout, saying, "Up and at it. Time

to strike while the iron is hot before the wind comes up." That was 4:30 a.m., and we were on our way at 5:15. We had the entire Audubon flotilla engaged, the *Puffin III, Osprey III,* and *Lunda.* We had packed the boats the night before, loading all of the pieces for the Egg Rock Hilton, as it came to be known. Joe took the lead at the wheel of the *Puffin III* and towed his trusty twenty-three-foot wood dory. John Federico, a large, always friendly, and competent captain, was at the helm of *Osprey III.* It took both boats to hold the deck, walls, and roof for the cabin. Joining me for this latest adventure were Evie, Rich Podolsky, Tom French, and Dave Enstrom, a new member of the crew. Dave had a passion for birds but also some skill at laying bricks, so we named him construction manager, a well-deserved title that he would earn over the next few days.

Never short of new projects, we also began a storm-petrel attraction project that year at nearby Old Hump Ledge that involved installing speakers and a battery-powered sound system. Our first stop was at Old Hump Ledge to drop off three heavy car batteries and a large box to hold the sound system. We soon arrived at Egg Rock, and Joe rowed the cabin ashore in eight full loads with the dory using his "Norwegian steam," which was always in good supply, especially in these early hours of the day. By 10:00 a.m., all of the cabin and personal gear was ashore, and as predicted the wind was "airing up" as Joe and John headed back to Hog Island, leaving us on the island with a huge load of lumber and the *Lunda* secure on her mooring.

I picked the site for the cabin on a high spot that was closer to the landing rocks than our old tent camp. The tent site that we occupied for the past five years was located near the puffin-rearing burrows, far from the landing, which meant we had to lug the water and batteries over lots of rocks to the old campsite. The new location offered a few more feet of elevation and a granite ledge that I hoped would help to keep the cabin secure against the inevitable gale-force winds. Using hand shovels and muscle, we dug holes down to the bedrock where the four corners of the cabin would eventually stand. We then filled the holes with the largest boulders that we could move to create piers on which we built the foundation. Somehow, with only a crude sketch, Dave figured out what Joe had in mind by looking at the numbers marked on the many pieces of wood that were now stacked up at the cabin site.

At the end of the first day, the floor, walls, and rafters for the shed roof were in place and braced. But rain had started in the afternoon, soaking us as we worked, and it kept coming down through the night, threatening to flood our tents, which we had set up near the cabin. The rain continued in a downpour the next day, but we kept working, and, by end of day, the walls and roof planks were in place. The rolled roofing, however, was not yet installed, and the rain leaked through, forcing us to continue sleeping in our flooded tent site.

By May 25, we were all so wet and exhausted that we decided to head back to the mainland for some dry clothes and a hot meal. Had we known what lay ahead, we would have stayed on the island.

At 8:30 a.m., I launched the *Geezer* and rowed out to the *Lunda*. During that short passage, the wind picked up to about twenty-five miles an hour from the northeast and became so fierce that it was all I could do to make it to the mooring. The thought of making three or four more trips in the *Geezer* was so daunting that I decided to free the *Lunda* from the mooring, motor in closer to the island, and drop the anchor to reduce the rowing distance. Not a good idea. Alone on the boat I was not quick enough to drop the anchor and steer clear of the anchor line. The propeller became entangled and promptly made the stern swing into the wind.

To my horror, water was now pouring over the transom, filling up the boat. I could not move with the motor so entangled, and there was no way that I could reach the propeller in the pitching sea. I was so close to the rocks that even if I had freed the prop, the wind and seas would have smashed the boat against the granite. It appeared that the *Lunda* was doomed, so I dropped the *Geezer* into the water and paddled ashore, leaving our precious vessel to founder.

Rich, Dave, Tom, and Evie were watching aghast from the landing rocks. Once I was back ashore, we discussed what to do. Tom, by far the strongest rower among us, agreed to row out to the *Lunda* and tie a line to her in case the anchor broke loose, but the wind was so strong that he could not reach the boat, which was now lurching wildly in the white water.

We raced back to the cabin to call Joe Johansen. We set up the CB radio, broke into a conversation, and asked for someone to call Joe on the landline. The next thing we knew, Joe's wife, Mary, was on the phone saying that Joe and John Federico were on the way.

The north wind must have helped Joe, because he made surprisingly good time reaching the island. On arrival, he was all business. First order was to get us off the island and onto the *Puffin III*, now manned by Federico. Joe rowed his dory into the relative lee on the western side of the island, shouted for us to leap aboard, and then mightily rowed back to the *Puffin* and let us off. From there, we watched the rest of the salvage operation.

With Tom still aboard the dory, Joe guided it back to the *Lunda*, which was still pitching wildly and now just a few feet from the rocks since the tide had dropped, exposing even more granite. Tom tied a line to the bow of the *Lunda* and then cut the anchor loose. Using only his "Norwegian steam," he towed the crippled *Lunda* away from the rocks and back to the *Puffin III* so that we could tow her back to the mainland. On our return to the mainland, Mary had bowls of hot stew waiting for everyone.

We called our new cabin the Egg Rock Hilton because of its relatively luxurious accommodations compared to the wall tent and because of the omnipresent "bellboy" that was always a resonant chiming background to our conversations. The Hilton was a humble outpost, fortlike, tucked against rock, and painted granite gray. The space was efficient like a ship's cabin, with kitchen, work area, lounge, and sleeping area arranged into one small twelve-by-twelve-foot space. The sleeping space was a loft just eighteen inches high, large enough for four assistants who would sleep sardine-like. With the roof just inches above the sleepers, there was no room for sitting up in bed. The cabin was conceived as a bird blind, with a hatch from the common space that opened to an observation structure on the roof. Like the remote Cape Cod cabin in Henry Beston's classic, *The Outermost House,* our "Hilton" became the outermost house on Muscongus Bay.

Since construction, it has survived several tropical storms and near hurricanes, with sea flotsam sometimes delivered to its doorstep during winter storms. Storm-petrels nest under it, and several hundred Project Puffin interns have slept on the roof and learned to care about seabirds and the sea, sheltered by its thin but persevering walls.

The first half of the 1979 summer continued to be promising but still a tease. We placed puffin and tern decoys back on the rocks on May 28. That day, terns landed, seeming to take

The twelve-foot-by-twelve-foot Egg Rock "Hilton" was built in 1979. It continues to serve as shelter, kitchen, lounge, office, and observation blind. Photo by Stephen W. Kress

interest in the decoys, often with fish in their beaks. Before the week was out, terns would come in and hover about a dozen at a time.

Not yet accustomed to sleeping in the Hilton loft, I nearly fell out of it at dawn on June 4 when new research assistant John Guarnaccia, a birding friend from Ithaca, shouted, "Puffin! Puffin!" The first of the season, the bird allowed us to observe it for nearly two hours.

Later in the day we talked on the CB radio with Kathy and Anne Hallowell, who were staying on Matinicus Rock as the guests of Carl and Harriet Buchheister, who were continuing their long tradition of spending several weeks there each summen helping to protect the seabirds and studying storm-petrels. Kathy and Anne were amazed to identify eleven translocated puffins among the native Matinicus puffins. They even reported a pair of Egg Rock–translocated puffins rubbing bills, the endearing behavior that occurs between mated pairs. Did this mean that our translocated puffins from Egg Rock had found each other and a nesting site at Matinicus Rock?

John spotted another puffin on the morning of June 9 near Egg Rock, the day *Wild Kingdom*'s Rod Allen, Marlin Perkins, and Ace Moore came to film our project. But the trio did not get there in time to see it, and they were shut out from seeing a single puffin during their five-day stay. I was concerned that the message of their film might be one of failure rather than success. My mood was not improved later that day when I ran the *Lunda* over her mooring line, slicing it off. But then I saw two puffins and a razorbill circle the island. The puffins had a way of showing up when I needed them.

Always looking for ways to tweak our attraction program, I added a four-sided mirror box to the cluster of puffin decoys atop Gull Rock. The idea came from my memory of keeping a mirror in the cage with a pet parakeet (budgerigar) in my childhood home in Columbus. I thought the visiting puffins might interact with their reflection in the mirror and stay longer, thus improving the chance that they would meet other puffins. As I hoped, the mirror was like an interactive decoy. The puffins responded in a similar way to my budgie of long ago. Puffins liked Gull Rock because it was near the edge of the water and the highest point on the island, giving the birds an easy launching place should they need to make a quick retreat from a gull attack. When puffins landed by the mirror, they tended to walk around the box, peck at it, and often sit down next to it for hours at a time.

I hoped the mirrors would keep the puffins on the island longer, thereby attracting other puffins. There was nothing notable over the next two and a half weeks, until July 4, when one or more puffins were spotted for five consecutive days. Egg Rock was starting to feel like a colony, but we still had no evidence of nesting.

Then another new record was set: a triple puffin sighting that continued for four days. New records were set on each of the next days with as many as five puffins at a time. We were on a roll, but still most puffins did not stay for long. We feared that the allure of Matinicus Rock was all-powerful. We had come so far, but six years into the project, we had not seen puffins going under the boulders or revisiting their original sod homes.

A soggy fog returned, and only single puffins were visiting the island when the time arrived for our return to Great Island with the *Wild Kingdom* crew. On this trip we accidentally collected an extra bird to make it 101, but the trip home to Maine sadly proved fatal to 3 chicks. Another died in its burrow a day after arrival. Still another perished the next day and then a sixth. The next day, a chick went missing. "Quite depressing and too reminiscent of some of last year's mishaps," Tom Fleischner wrote. "I am completely baffled as to what happened."

Then Tom confessed to reaching "critical rock fever." He asked for a break off the island because the "overbearing weight of responsibility for each bird's welfare is incredibly time-consuming and physically/emotionally exhausting."

On the brighter side, we had tallied twenty-one consecutive days of seeing at least one puffin and sometimes as many as seven. Enthusiasm quickly replaced despair. The next day: ten puffins, the first double-digit day since the late nineteenth century, Marlin Perkins and the *Wild Kingdom* crew were back on Egg Rock and happy to be part of the excitement. One of the highlights of their film featured me sitting next to Marlin in the weatherbeaten Flagpole Hill bird blind, where I showed him several puffins parading among the decoys. Although the film was largely staged and rehearsed, the puffin appearance that day was genuine.

Marlin left a special message in the Egg Rock journal: "The puffin transplant is indeed a wonderful story and evidence has come in to show that it is working . . . Praise the Lord."

The beginning of August brought what had become an annual ritual: removing the wire mesh that restrained the pufflings and replacing this with dowel activity indicators. The anticipation surrounding the exodus of chicks was sharply heightened by the returning puffins. Two mornings after the sighting of ten, eight puffins clustered together around the mirrors and decoys on Gull Rock. That evening, an eleventh puffin came onto the island.

When Evie and I watched a group of six puffins waddle on the rocks, I said to her, "Do you realize where those birds have been? They were in my car!"

One more chick died during all this, bringing us down to ninety-three chicks, but the sorrow did not drown out the new wonders we were seeing as a raft of twelve puffins swam around Egg Rock on August 3.

The wonders kept coming even as our latest chicks fledged. On a final puffin-watching stint in August, Evie, Richard, and Diane DeLuca saw a puffin land on Gull Rock with a face full of fish! A puffin with fish usually means a chick feeding, but not always. We marveled for a moment, but the puffin flew off with the fish rather than dropping into a rock crevice. The sighting became yet another mystery.

Then all life and natural phenomena picked up. By August 21, Diane and Tom saw nine puffins circle the island, as if giving the team a grand send-off for the season. Every part of the seascape was dotted with feathered reminders of southward migration: loons were now moving south in flocks, and hundreds of ruddy turnstones were roosting along the shore with a few red knots and sanderlings. On the island, northern waterthrush were creeping through the raspberries, and red-breasted nuthatches were clinging to the granite. Even ruby-throated hummingbirds would occasionally buzz the island. One day, Diane and Tom caught in the mist net a hissing and clicking red bat, yet another migratory surprise. One night the sky was lit up with stars and waves of aurora borealis. At dawn, low tides exposed starfish a foot in diameter. By the 27th, all but two chicks had fledged, and three adult puffins still frequented the island. The only sober moment of these electric days was Rich's discovery of the decomposed body of the chick that went missing early in the summer.

On the morning of August 29, as Evie, Rich, Diane, Tom, and I stood on the shore, a puffin flew right past us. It was likely the last little brother of the season.

Two days later, I brought my parents to Eastern Egg Rock for a visit.

It had been many years since my mother dropped me off at the Greenwich Audubon camp and told Peggy Morton about my boyhood of bringing home frogs, salamanders, and snakes. I was thankful that it was a calm day. My parents came ashore, and I was able to show them the Hilton and a bit of the excitement of the project. They would have been concerned to see the raging shore that we sometimes experienced and to sense the dangerous side of the project. It was a foreign experience to them, but they could see my passion for what I was doing and understood that this was my world and they were proud to be part of it.

On September 4, we made one last visit for the season to tackle one last session of burrow repairs. It was satisfying to be at the island without the worries of looking after the birds and to know that we had managed to bring another hundred chicks to Maine and nearly all of them had left the island as robust, healthy puffins. Now it was up to the birds to survive at sea and, we hoped, reclaim these long-silent boulders.

Yet my concern was growing. We were seeing more puffins, but they were not going under the rocks. Perhaps they were drawn back to Egg Rock as their homeland, retracing some magnetic map, but without established breeders, would they nest? I knew that many were also visiting Matinicus Rock, though I couldn't be sure it was the same individuals frequenting both islands. We had come so far, yet the outcome of the project was still out of reach.

The following summer—1980—puffins were not yet nesting, but researchers were. On April 27, newlyweds Tom French and Kathy Blanchard-French were back on the Rock. Kathy and I had once shared the puffin dream but found it difficult to move our relationship forward. Now she and Tom were a pair, as were Evie and I. We were here to see how the Hilton had fared over the winter. Other than having a drooping rain gutter, the Hilton was fine. We installed four standing puffin decoys out on Gull Rock and put out thirty-three tern decoys in the mock colony.

We left the island in isolation for a month until May 25, when Joe and Mary Johansen brought Evie and me out to Egg Rock to set up camp for the 1980 season. As Joe moored the *Osprey III,* a puffin whipped out of the fog as if to greet us. Five puffins were already on the island—the first time the little brothers beat us to the rock! By June 2, when Tom French, Rich, and research assistant Mark Jaunzems landed for a stint, there were nine puffins on the island.

On June 4, Mark saw two puffins mating on the water and shouted in pen in the island journal, "It's a COPULATION!!" On June 5, a new record of sixteen puffins were seen on Egg Rock, with one of them carrying a blade of grass in its beak, a hopeful sign of nest building. On the seventh, Rich wrote that the burgeoning puffins "broke up into three groups of three-to-four birds each and began to actively court on the water." Between 6:00 a.m. and 11:00 a.m. he witnessed four copulations, three among pairs of puffins and one between a puffin and a lobster buoy.

That day marked the first time we observed a puffin drop down into the crevices underneath the boulders. "Today was the first real investigation of burrows," Rich wrote, with as much hope in his pen as actual proof that this was the puffin's intention. He reported that the bird was a yellow-banded puffin. It had jumped into a crevice and was out of sight for a full thirty seconds. This was huge news. He also noted that nearby guillemots were upset having their big-beaked cousin visiting in this manner. Such puffin explorations often ended with a guillemot chasing the puffin far from the island.

On June 24, Diane DeLuca was back on the island. Shortly after 6:00 a.m. she shouted, "Evie, nineteen puffins!" Just as stunning, at least one puffin from each of the four age classes was in the gathering. On June 26, I read the band number on a puffin that we had brought down from Great Island in 1978 and had seen as a one-year-old at Matinicus Rock in 1979. Here it was back at Egg Rock in 1980. Now we had proof that the puffins were moving between Matinicus Rock and Egg Rock. This puffin was observed for only a few minutes, emphasizing how easy it would be to miss such important sightings. The puffin data were coming in fast, so we took the bold move of building burlap-covered observation blinds and setting them in the puffin-nesting habitat.

The puffin increase paralleled a relatively explosive gathering of terns. I wondered if the increasing tern activity was helping to attract the puffins, which typically nest with the feisty terns that help chase off gulls and other predators. By late June, tern numbers were at recent record highs, sparked by a combination of gull control, decoys, and audio recordings of tern calls. We continued as we had the previous year, placing the decoys in the same location, but we decided to keep the audio running from dawn to dark, because we had earlier noticed that terns were most abundant when the "comfort sounds" were playing.

At this early stage of recolonization, the mirror box was especially appealing. One love-struck Arctic tern spent four straight days spellbound in front of the mirror on Gull Rock, demonstrating behaviors typical of normal courtship encounters.

On June 22, this tern lifted off from the mirror, went out to sea, and returned with a fish in its beak. It brought the fish to the mirror and paraded around the box with tail raised and wings dropped, vocalizing the soft keh-ke-ke-ke-ke-keh sounds typical of a fish-offering toward a mate. This being a no-win situation, the bird flew off and returned less than a minute later, minus fish. Again landing next to the mirror, the bird began strutting with tail raised and shoulders nearly scraping the rock, shaking its head side to side. Raucous calls from the northeast announced the arrival of eight more Arctic terns and four common terns. The mirror bird flew up to join in a twenty-five-minute caucus of frequent landings, displays, and chases around the south end of the island.

By July 5, about a hundred terns had settled on the rocks where we had arranged the decoys. Here we noted frequent copulations, and certain pairs became extremely aggressive toward other terns. Given how late it was in the season, we wondered if the birds had failed elsewhere and were making a second nesting attempt at Egg Rock. As expected, the terns showed no aggression to the occasional puffin they encountered.

Then a few roseate terns showed up, a new species for the project. We were hopeful that Maine's rarest tern would someday nest here. After ten days of these mass gatherings, I brought the team together on July 7 to inspect the hub of tern activity to see if it was still a mock colony or the real thing.

Evie, Diane, Dave Enstrom, and I went out into the grass tufts, spacing ourselves at arm's length, walking very carefully. I let out a whoop on discovering the first nest, a well-built cup of vegetation holding three eggs, presumably those of a common tern. Soon everyone spotted nests.

The terns hovered overhead but did not attack us. We found sixty-one glorious speckled beige- and olive-colored eggs in thirty-two nests.

The idea that we could transition from decoys to a robust colony in just a few days speaks to the need for the birds to nest together and share synchrony in hatching. One of the benefits of colonial nesting is to "swamp" predators by having many aggressive adults nesting near one another. Synchrony leads to many eggs and chicks present at once, which reduces the risk at any one nest. It's all about "safety in numbers."

I was thrilled to know that we had tilted events to favor the terns. This was the first nesting of terns at Egg Rock since 1936, a hiatus of forty-four years. But even more significant, this discovery marked a major event for seabird conservation. This was the first nesting of seabirds anywhere in response to

Arctic terns nested among decoys and audio speakers in 1980—the first response of a seabird to "social attraction." Here a common tern inspects an Arctic tern decoy carved by Donal O'Brien. Photo by Stephen W. Kress

the use of decoys and audio recordings. The method came to be known as "social attraction," and we were enjoying the first proof of success.[1]

By mid-July, puffin numbers reached twenty for the first time, and the ground count of terns hit nearly 150. On July 20, we tiptoed again to the tern meadow. One of the eggs had hatched. A tiny fluffy miracle, an Arctic tern chick was alive and well at Egg Rock. We also discovered that the colony had doubled in size, with seventy-two nests containing 170 eggs. The discovery of the first hatched tern coincided with the arrival of our next hundred puffin chicks from Great Island. The collection itself went off as planned, but we were reminded of the extreme danger of seabird work.

Just days before Evie and I arrived in Newfoundland, a

The tragic death of a seabird researcher landing at Great Island, Newfoundland, in 1980 forced the collecting team to land at the backside of the
island—a much longer trek with the heavy carrying cases to reach puffin
nesting habitat. Photo by Stephen W. Kress

boat of seabird researchers capsized at the Great Island landing, and a researcher with whom David Nettleship had worked
for seven years drowned. David had called us to cancel the
collecting trip, but we were already under way and didn't hear
news of the accident until we arrived at John Reddick's home
in East Bauline. We avoided the site of the accident and approached the island from the lee side, which meant a much
longer and steeper hike up the back of the island. On the way
out, Evie overheard John Reddick reflect on the accident, "The
sea forgives no one for a mistake."

By early August, adult puffin numbers reached a new high
of twenty-three. We busied ourselves with banding tern chicks
but wondered how many of the still-downy chicks would be

old enough to follow their parents to the southern hemisphere when the migration urge would lure the adults southward.

Although there was much to celebrate this summer, especially with the terns nesting, I knew that critics could quote our own statistics: seven years into the project we had now fledged 630 puffin chicks into the wild, and though many had returned and some were five years old, there was still no breeding.

Not on Egg Rock anyway. Probably the most exciting and nettlesome report of the summer of 1980 came on July 19 from Matinicus Rock. Here, Dave Enstrom reported that one of our Egg Rock "graduates" delivered food into a nesting burrow at Matinicus Rock. He noted that the puffin wore a single white band, just like the one that dazzled me at Egg Rock three years earlier. Perhaps it was this same puffin now calling Matinicus Rock its home. Although it was the first known breeding of a transplanted puffin and it had nested in Maine rather than Newfoundland, I was less than jubilant because it nested at Matinicus Rock rather than Eastern Egg.

One question gnawed at me: Was the existence of a nearby puffin colony undermining our effort to start a new colony and expand the puffin's range? Would all of the other translocated puffins make the same decision to settle at Matinicus Rock or some other established puffin nesting colony? Once they chose a nesting island, I knew they would stay there for their long reproductive life. That fact would either undo everything or someday help to secure a breeding population on Egg Rock. The encouraging news was that the puffins that were returning to Egg Rock seemed to be increasing their burrow investigation. Our goal seemed closer every day.

Yet where were the hundreds of puffins that we had moved from Newfoundland? Most were missing. We were either at the brink of triumph or tragedy.

9

Puffin . . . with Fish!

Aside from the biological realities about the outcome of the project, I had plenty of reasons to wonder how long I could keep the puffin project going. I knew I couldn't continue to rely on the small year-to-year research grants from Audubon for a long-term program, and I also realized that skeptics of the project were waiting to claim victory if the puffins didn't nest soon.

I hoped a grant of four hundred thousand dollars in 1980 from the Celanese Corporation would mark the beginning of serious fund-raising by Audubon to further the program. The grant was intended to support Audubon's work on behalf of California condors, whooping cranes, bald eagles, and puffins. Eyebrows were raised, however, because Celanese was a titan in the chemical and plastics industry, ranked 107 in the Fortune 500 that year. Biologists were realizing that the synthetic world, once thought to be a modern miracle of comfort and convenience, was becoming a disaster in the natural world, and its negative impacts on ocean life were becoming increasingly conspicuous.

Fishing lines and six-pack rings were obvious hazards,

snagging and choking seabirds, marine mammals, and sea turtles. Plastic trash had come to make up nearly 90 percent of the ocean's garbage, so omnipresent that breeding seabirds were gobbling up everything from bottle caps to grocery bags. Even party balloons from Midwestern parties were finding their way to the oceans, choking seabirds and other marine animals. It was also increasingly clear that plastic does not go away—it just gets smaller and smaller, creating a cascade of trouble deep down into the marine food chain. Degraded bits can be picked up by puffins as pseudo-marine worms, and even smaller bits and plastic pellets can concentrate toxic chemicals such as PCBs and DDE before they are consumed by fish and fish-eating birds.[1]

Seabird chicks such as Laysan albatross are especially vulnerable when parents bring back hard plastics such as cigarette lighters that the chicks cannot cough up, eventually leading to starvation.[2] By the mid-1980s, National Audubon's concern for birds was joined on other fronts. A consortium of fourteen environmental and marine wildlife conservation groups called the Entanglement Network joined forces showing outrage about plastics in the ocean. The consortium estimated that as many as two million birds, including puffins, terns, albatrosses, murres, petrels, and gulls, and at least one hundred thousand sea mammals, such as whales, seals, manatees, and dolphins, were dying annually from plastic choking or ingestion.[3] There is so much trash in the sea that every summer when we go back to Eastern Egg Rock, one of the first tasks of the season is clearing heaps of Styrofoam flotsam, plastic bottles, and tangles of anything that can choke the birds.

Celanese—or any plastics company, for that matter—was never linked directly to the trash or to the specific chemicals such as DDT, DDE, and PCBs that poisoned eagles and many other birds and their prey, but lobbyists from chemical, oil, and metal industry giants were working to weaken pollution regulation under the Clean Air Act. So it was no surprise that a *Sports Illustrated* article greeted the Celanese gift with the headline, "Strange Bedfellows," calling the gift "an unusual alliance of corporate and conservation interests."[4]

This was a moment of an unprecedented nexus between National Audubon and many other environmental groups, such as the Sierra Club and the Wilderness Society. Citizen memberships were growing dramatically as a result of the awareness spawned by *Silent Spring* in the 1960s, the first Earth Day in the 1970s, and the anti-environmental policies of the newly elected Reagan administration. But precisely because of that awareness, nimble corporations were clearly trying to stay out of condemnation's way. The conservative Heritage Foundation quoted a 1980 report by Environmental Action that found "an explosion of interest in corporate giving among many environmental groups," including Celanese's four hundred thousand dollars to Audubon.[5] In the middle was National Audubon's new president, Russell Peterson, who took a circuitous route to the executive post of a national environmental organization. A 1942 PhD chemistry graduate of the University of Wisconsin, he made a career at DuPont.

In 1979, National Audubon, impressed with Peterson's straight shooting, hired him to come to the other side of the environmental debate. Then came the election of Ronald Reagan in 1980 and a doomsday scenario of cabinet appointees who appeared to wish to hand over stewardship of the nation's coastlines to industrial interests. How far the administration wanted to turn back the clock was clearly spelled out in its first budget proposals, with a massive one-billion-dollar cut to the Department of the Interior. The US Fish and Wildlife Service, an agency within the Department of the Interior, owned and managed many islands off the coast of Maine, and it was my primary partner in Maine seabird conservation.

Peterson's mantra was "Think Globally, Act Locally." Audubon was becoming global in its vision, from a concern about nuclear power to proposing raising fuel economy from 27.5 miles per gallon to 37 miles per gallon by 1990 to government subsidies for Americans to install rooftop solar panels.

I agreed completely with Peterson's philosophy, but I wondered what Audubon's public thrust into national politics and the growth of corporate donors meant for Project Puffin. Peterson shook up the Audubon administration soon after coming onboard and in the process sacrificed much of the organization's institutional memory. Sadly, the regrouping meant that my mentor, Duryea Morton, resigned his position as vice president for education. With Morton's departure, the Audubon camps lost their greatest friend, breaking the direct link to Carl Buchheister. In the process, wildlife sanctuaries and environmental education took a back seat to public policy. As for the big Celanese grant, I never saw any of it directly benefit Project Puffin.

My winters would continue focused mainly on raising private funds for Project Puffin. I was donating my time, and the interns were paid a meager stipend for the summer. A small group of loyal supporters kept the project going, but they were starting to ask questions about when puffin nesting would occur.

By the summer of 1981, Peterson was so angry about the Republican Party's efforts to gut environmental laws that the *Christian Science Monitor* quoted him as exhorting at an Audubon meeting in Rocky Mountain National Park in Colorado, "The Reagan administration is attempting a coup—attempting through administrative actions to circumvent the environmental laws of the land, and to turn our natural resources over to the exploiters—to the modern-day plume hunters and buffalo hunters. We must stop them."[6]

For instance, when the California condor reached the brink of extinction in 1979, Audubon pleaded with the US Fish and Wildlife Service for assistance. Audubon chronicler Frank Graham Jr. wrote that Fish and Wildlife believed the bird to be doomed, so if Audubon wanted to jump-start a salvation project, it first had to put cash in the game. Audubon said it would make three hundred thousand dollars available.

Because the puffin was not endangered, I wondered where my bird stood when it came to dividing up grants such as the Celanese gift. I had argued that the time to develop restoration methods was when species were still abundant, but this was a hard case to make when limited dollars within Audubon were being focused on endangered species. It was no real surprise to me when the Celanese funds shifted to endangered species and I was on my own raising dollars for puffins.

I had two immediate concerns. One was whether I would be able to procure more chicks. Seven years after the project started, I was still waiting for our first breeding puffin. I was convinced that the increasing number of sightings and that most of the puffins were still too young to return positioned us well for success, but it was still a hunch and the naysayers were at my back. Clearly, the importance of Project Puffin was not obvious to most, especially given the national background politics that were so weighted against conservation.

Second, I was surprised to find that Ralph Palmer had resurfaced and become a nemesis for the project. He not only was a cynic about the value of puffin restoration but was quietly fuming about the publicity received in the press about Project Puffin in recent years. I had not heard directly from him since he denounced my original idea nearly a decade before. Now he pounced on the announcement of the Celanese grant as evidence that Audubon was using puffins as a public relations ploy for fund-raising. He had been fuming about the project since its inception. Now he was insisting to whoever would listen that Project Puffin was no more than a public relations stunt.

Palmer knew Roland Clement, at the time National Audubon's vice president for science, and he took the opportunity to bend Clement's ear about the project. Palmer also wrote in the summer of 1980 to Anthony Erskine, the chief of migratory birds management for the Canadian Wildlife Service and Nettleship's supervisor, complaining that people were disgusted with the ballyhoo about puffins. He also continued to express doubts that puffins would ever breed at Egg Rock. Erskine responded to Palmer that publicity should not be dismissed as irrelevant, but he agreed that it was important not to lose focus on other wildlife challenges while focusing on puffins.

Clement decided not to take Palmer seriously and understood that the reason it was okay to restore ranges for still common birds in the United States was exactly the reason that bothered Palmer. Because puffins were still abundant elsewhere, experiments with a few puffins at the edge of the range would not affect puffins where they were still abundant in places such as Newfoundland and Iceland. Eventually Clement decided that the outcome of Project Puffin would not be determined overnight but that ultimately it was best to let the puffins decide if they would call Egg Rock home.

The Fourth of July festivities in Round Pond, Maine, in 1981 included the usual parade featuring ancient cars, marching bands, and quirky floats poking fun at politicians and tourists. It was not especially memorable as Round Pond festivities go. But this Fourth of July changed my life. I was just a few miles away, but I might as well have been on a different continent. The way I looked and smelled would have made me an oddity on the mainland, where townsfolk counted down the hours to fireworks and tourists dressed in red, white, and blue swarmed the famous gnarly rocks of nearby Pemaquid Point.

Evie Weinstein and I were perched on opposite ends of Eastern Egg Rock, a seven-acre, state-owned, treeless island six miles off the mainland. I was scrunched into a tiny, three-foot-square plywood bird blind, confined like a stuntman in a barrel about to career over Niagara Falls.

Because we knew about all the festivities, this seemed to be an especially lonely day. A pea-soup fog cloaked the rock, preventing us from seeing the mainland and erasing any chance of seeing fireworks. Even the lobster boats that normally churned past intermittently were absent. The men and women who haul up traps containing Maine's signature delicacy were enjoying their own clambakes along the coast, dropping fringe benefits into bubbling cauldrons.

I was in a blind we nicknamed Flagpole, because near this location we hoisted a puffin flag in the earliest days of the project. Evie was at the Egg Rock Hilton catching up on journal notes before taking my place in the blind. As those on the mainland celebrated the independence of a nation symbolized by the bald eagle, we were in a silent vigil, hoping to see puffins.

I had sat in the blind for five hours, from nine in the morning to two in the afternoon, and I was ready for a break. All I had to show for my time was strained eyes. Nothing emerged from the fog. As if to amplify that this was another fruitless stint, rain and a chilling wind blew in from the northeast just as I tucked my weathered seat cushion into the bucket that sufficed for my chair in the blind. I bent low to slip out the door of the blind and worked my way over the boulders toward the Hilton.

I was thinking about what canned food to throw together for a late lunch when Evie exploded out of the fog toward me from her end of the island. She was waving her arms in the contortions of a cartoon traffic cop signaling everyone to proceed simultaneously in all directions. Her whirling limbs prevented her from talking, let alone shouting. What could this be about, on this otherwise grim, almost lifeless day? What could she have seen that shattered the tedium, shocked her into pandemonium?

"Puffin . . . with fish!!!" she finally blurted out.

"Where!?!?!? When!?!?!?" I stammered.

Breathless with excitement, Evie explained that she had been collecting seawater for washing dishes when—like an apparition—a puffin emerged out of the fog and buzzed past her carrying the loveliest fish she had ever seen! She noted with precision in her field journal that this had occurred three and a half hours before, at 10:36 a.m. She went on to explain that the puffin was flying directly toward my blind.

The sight of the puffin with fish draped out of its beak was so surprising that she had dropped the bucket of water and was torn between dashing to tell me and not wanting to disturb my stint or the puffin's flight. So she raced back to our tiny camp with the urge to tell someone about her most amazing vision! She had thought to wave down the first lobster boat to come by, but they were not to be found on this foggy holiday.

Desperate to share her news, she called the Camden marine radio operator, Marge. Most conversations with marine radio operators, unless they involve true emergencies, are routine, almost military check-ins seeking a telephone communication. They are not small talk.

Evie broke protocol because the news was so big. Later she recalled, "It just seemed so impossible, I had to tell SOME-

BODY! I was so hysterical that when I got Marge on the radio, she asked if it was an emergency. I said 'No, no, no!' and then I launched into one-word sentences: 'Puffinwithfishinitsbeak! PuffinbabiesonEasternEggRock! Igottatellsomeonesolcalled youfirst!'"

Marge responded patiently, "That must have been cool." After she ended the call with Marge, Evie said, "After I got off with Marge, I was hopping around like a goofy kid."

We rushed back to the cabin to focus our Questar on the south end of the island. This was a powerful birding scope on loan for the summer, a far more precise tool than binoculars. Evie was absolutely sure of what she saw, but several hours passed with no new sightings. Afternoon became evening. Then at 7:40 p.m., with the Questar scanning the south end of the rock, a puffin sliced out of the fog with its beak packed full of silvery fish. It scrambled over the rocks, then disappeared into a crevice! Fifteen minutes later, it popped up from the crevice, with no fish.

We watched in wonder as it hopped onto the rocks and paused near another puffin that appeared to be standing guard. Perhaps it was a mate. The puffin that had come with the fish flew away, most likely to find another meal. One more time, twenty-five minutes after that, as daytime gray faded to night black, the feeding puffin repeated the performance, delivering more fish. We were beside ourselves with joy! The delivery of fish meant only one thing—there was certainly a tiny, fuzzy chick somewhere deep in the granite boulders.

While the faint boom of fireworks cut through the fog, I reflected on the past hours. This was the day we had experienced a thousand times by dream, but seldom mentioned, for fear it would somehow jinx the experiment. We had seen a puffin with fish in its beak buzz Eastern Egg Rock two years before, with no proof it ever landed, let alone had a chick. This time though, I was amazed to see a puffin with its beak crammed full of fish enter the field of my scope and walk-fly-scramble into the boulders below. This was the best proof we could have that after almost a hundred years of absence and eight years of working toward this goal, puffins were again nesting on Eastern Egg Rock. A Fourth of July celebration I would never forget.

Evie's three words, "Puffin . . . with fish!!!" were the three most important words of my career. That Atlantic puffin, with those fish, represented something that had never been accomplished before. For the first time, a seabird was restored to an island where humans had wiped it out.

On July 13, nine days after Evie spotted our first incoming puffin with fish for a chick, I attempted one last response to Ralph Palmer and did my best to remain civil: "I regret that you do not see merit in our seabird re-establishment efforts. Regardless, I'll keep you advised of the progress of our work with the hope that you will eventually see value in it. Even though I find your comments unduly critical and suspicious of our motives, I do find your views an instructive balance to what I see as generally very strong support for our accomplishments. I have not dismissed your comments and recognize that they come from a sincere concern for Maine seabirds."

A month later, Palmer sent a highly sarcastic and bitter recognition that puffins were again breeding on Egg Rock, saying that he hoped the project would succeed vastly so that there would be plenty of "stewed puffin." He went on to say that nothing altered his view that a seabird sanctuary meant leaving the birds alone and that "tinkering with puffins" at the edge of their range was of value mainly to the media.

After eight years of waiting for puffins to nest at Eastern Egg Rock, the sighting of a puffin dropping into the rocks with fish offered proof that puffins were nesting again at Egg Rock after an absence of nearly one hundred years. Photo by Derrick Z. Jackson

Fortunately the popular press remained favorable to my project throughout its first seven years, despite not yet having breeding birds. A short article by the *New York Times* after the 1980 season described me as saying that I believed the project was "nearing success."[7] That season, I showed a *Christian Science Monitor* reporter a transplanted chick and optimistically said, "See his little wing feathers forming? They'll be powering him around the island someday."[8] When we announced the first breeding puffins in 1981, *Newsweek* proclaimed, "The puffins had finally decided to call the Rock home."[9] These early results and subsequent events were also published in several scientific papers.[10]

By the end of 1981, we tallied five breeding pairs, including one pair without bands that puzzled us, but we were happy to have these mystery birds join the colony. The other eight were all Newfoundland puffins transplanted by airplane, boat, and Johansen's Norwegian steam as fluffy pufflings. Now they were nesting, and we could not have been prouder. Most of these founders returned in 1982, and the colony increased to fourteen pairs—accompanied by more than four hundred pairs of terns. We reveled in this success, but I had no idea how much tinkering would be needed to move the restoration forward over the next three decades.

10

Roger's Rogue Wave

It took four years for the first puffins to come back to Egg Rock and eight years for the first breeding. Once they were established, we had hopes for an exponential explosion of birds. But a sober reality set in. The euphoria of the first breeding puffins on Egg Rock in nearly a century faded into feelings of vulnerability. For all of our intervention, the hand feedings, the decoys, the taped tern calls, and Gramlich's gull poison, the pace of restoration was similar to the slow progress of colony growth at Matinicus Rock, which was aided for decades merely by a vigilant armed lighthouse keeper.

At moments, the project almost felt stalled, and I began wondering if our accomplishments would begin to fade away. We still knew nothing about where our precious few puffins went for eight months of the year and had no idea what dangers they faced away from our protection. It seemed likely that all of our effort would be hard to sustain as long as the puffin numbers were so small, especially when one considered that one severe winter storm or raid by a prodigious predator could wipe out all our efforts.

My internal worries for the future grew as film crews became enthralled with the present. Our revitalized seabird colony and our humble home on this remote island were story enough for numerous articles and movies. We entertained visits by *CBS Sunday Morning,* PBS's *Newton's Apple,* and many other programs for short news pieces. A one-hour film produced for *National Geographic Explorer* in 1987 was for many years the best documentation of our work. The producers, Michael Male and Judy Fieth, were perfectionists. Unlike most of the news shows, which wanted to tell our story in a few minutes gleaned from one half-day visit, Mike and Judy dedicated a full summer to making their inspirational film.

I was pleased when they contacted me again in 1989 to tell me about a new, hourlong documentary they were producing on Roger Tory Peterson titled "Celebration of Birds" for the well-known PBS series *Nature.* They started filming at the Ding Darling Refuge in Florida in January 1990, and then they moved up the east coast, filming and interviewing Roger as he revisited important birding locations such as Delaware Bay, where they recounted the saga of the red knot and horseshoe crabs. Then they traveled to Roger's hometown of Old Lyme, Connecticut, where he narrated the demise and recovery of ospreys following the banning of DDT. They wanted to bring him to Maine by late July, where they would film Peterson's return to Eastern Egg Rock, where he had brought Hog Island campers on field trips more than fifty years earlier. The idea was to show the possibilities of seabird restoration and the progress we had achieved.

At eighty-one years of age and recovering from prostate surgery, Roger seemed frail. Our paths had crossed many times over the years since Irv Kassoy first introduced us in Columbus. Now he was decorated by presidents and widely recognized as the "father of modern bird watching." He walked with the aid of a monopod (which gave me pause as I considered the slippery landings and irregular terrain at Egg Rock), but he was undaunted by the less than ideal weather and the thick fog that greeted him at the top of the hill looking over Muscongus Bay. He was fascinated by butterflies, which were abundant at the Audubon property. So there was plenty to occupy him while he waited several days for the weather to clear for the visit to Egg Rock.

On Tuesday, July 24, the fog cleared, and we moved forward with the plan to take Roger, Mike, and Judy to Eastern Egg aboard my twenty-three-foot Seaway, the *Lunda III*. But the CB radio report from Barbara North, the island supervisor, said sea swells were high, which was problematic considering Roger's mobility and the heap of equipment necessary for filming and camping. While making arrangements for the landing, I remembered what Roger's wife, Ginny, had said to me, "Take care of Roger—he is a national treasure."

To help with the landings, I arranged for Joe Johansen to join us so that Joe could row Roger ashore in his twenty-one-foot wooden dory along with Mike and Judy's equipment, which included a fifty-thousand-dollar camera. After tents were up, Mike and Judy filmed an interview between Roger and me.

I headed back to the mainland to resume my responsibilities at Hog Island, leaving Roger to revel among the seabirds and entertain Barbara, intern Tracey Tennant, and Debbie Zombek, a volunteer from SeaWorld in Orlando, Florida, with his endless tales of travel and birding. Later, their journal entries would describe in detail how much fun they had with "The King Penguin."

Two days later, I went back to the island to retrieve Roger,

Mike, and Judy for a celebrity evening on Hog Island. We were
joined by Jerry Skinner, a marine life instructor on Hog Island, and Captain Bob Bowman, a noted Bar Harbor whale
researcher who happened to be visiting. Everyone wanted a
chance to spend a little time with Roger.

Ever since Marlin Perkins had refused to be photographed
in the *Geezer,* our orange "Sportyak," we had switched to using
rubber inflatables. Now we were going to count on the inflatable to carry Roger and all of his gear from Egg Rock back to
the *Lunda* for the trip home. We had timed the pickup for high
tide, which would make it easy for Roger to get into the rubber
landing boat and to climb aboard the *Lunda,* but that pushed
the trip into late afternoon. It was already about 4:00 p.m.
when we left Eastern Egg Rock. The afternoon sun had burned
off the fog, and our visibility was good. Roger made it into the
boat without much difficulty, the sea was calm, and we didn't
bother with life jackets.

On the way back to Hog Island, we detoured three miles
to the west to circle Western Egg Rock, the treeless, ten-acre
island that I had scouted for puffin nesting habitat two decades earlier. I had not visited it in many years, but I thought
that Western Egg was a good control, to show Roger what happens in the absence of a seabird restoration project. As we approached the south end of the island, we could see the nesting
gulls and cormorants. The island looked just as I remembered
it twenty years earlier.

Mike was filming Roger while Judy asked questions about
his memories of the place in the 1930s. I was doing my best to
position the boat so that the island and birds made a backdrop
behind Roger while keeping an eye on the waves, which had
kicked up a bit. When they completed filming, we circled to
the north end of the island.

"Would you like to make another pass around the island?" I asked.

The filming resumed as we began our second pass close along the eastern shore, where dozens of eiders were bobbing in the surf under a large granite knob. Then I brought the *Lunda* a second time around the south end, keeping her several hundred yards offshore, away from the rollers that were now pounding the shore. We were in about thirty feet of water. Mike and Judy continued to film and record, Judy with microphone in hand and Mike with his huge camera on his shoulder powered by the lead battery belt he wore tight to his waist.

I was the first to see the wave coming. A huge wall of green water was rushing at us. I shouted, "Here comes a big one!" In that blurry instant, I was looking up at an angry crest just before it curled down and crashed into the *Lunda* with such force that the port side of the boat where Michael was standing was pushed down, flipping the *Lunda* completely over. I recall thinking, "This can't be happening! This has never happened before!" But it was happening, and we were all thrown into the frigid water, with all of our equipment sinking and floating around us. Somehow my binoculars remained around my neck.

Mike went straight to the bottom, weighed down by his lead battery belt. Tangled in cables and still clutching his camera, he probably would not have made it back to the surface without Judy diving down to free him of the weight. The fifty-thousand-dollar camera was ruined. Miraculously, we were all near the boat and keeping our heads above water, terrified as more waves rolled over us, getting weaker by the instant as the fifty-degree water sapped our energy.

At first we couldn't find Roger, but much to our relief, he soon bobbed to the surface near the capsized boat. We mustored our strength to pull ourselves up onto the slick belly of the *Lunda*. Together, we hauled Roger out of the foaming water onto the boat, but before we could catch our breath, another great wave broke over us, washing us back into the frigid water. Now we were drifting closer to Western Egg Rock. I was afraid the boat might crush us against the granite, so we began swimming toward land. Eventually, we started finding welcome rock beneath our feet as we neared the shore.

Roger was completely spent. Bob Bowman, a sea captain and probably in the best shape of anyone in the group, did most of the heavy lifting to keep Roger afloat while everyone helped. Finally in the shallows, we dragged ourselves ashore over the rocks and slippery seaweed like so many drunks. We were soaked, exhausted, and chilled to the bone, but everyone seemed to be okay, except Roger. He was bleeding from barnacle scrapes and holding his ribs in pain. Hypothermia was setting in, and he was shivering. He could barely stagger forward with our help.

We laid Roger down, and Mike huddled next to him to share his body heat. The *Lunda,* still belly up, was drifting ashore on the high tide, its propeller out of the water. I knew that our luck was turning when I noticed that some of our equipment was floating ashore. The best prize was a tent, which we promptly set up to create a shelter and windbreak for Roger. The sun warmed the tent's interior, and we all took turns lying next to our birding hero, talking to him, afraid that if he fell asleep, he might not wake again. His quiet voice was rambling; I feared he was becoming delusional.

The butterflies were migrating that day. Even in his stupor, Roger noticed the red admirals and monarchs alighting on the tent, backlit by the afternoon sun. While I lay next to him, I heard him ask if these were angels. Much later, he re-

counted that he actually said they were angelwings (butterflies), which may have been the case, as I have seen this beautiful orange butterfly with mauve-edged wings in late summer
on these islands. But perhaps he did mean angels, for he went
on to tell me that while he was underwater, he thought that he
saw Mildred, his first wife. I later learned that his first wife had
met a tragic, mysterious end by drowning in this same bay.

Our fortune improved further when Bob Bowman discovered that his cigarette lighter could muster a few sparks.
Quickly, we began gathering combustibles. Three great piles of
driftwood, lobster buoys, and other junk were set ablaze. I was
never so pleased to find such abundant trash on a seabird island or to be in company of a smoker. The blaze grew large and
black with acrid smoke, which we hoped would attract the
attention of a passing boat. But it was now approaching 5:00
p.m.; lobster boats were long off the water, and no recreation
boats were passing within sight.

It was getting dark. I knew that as the air temperature
dropped, the fog would return and our chance of rescue would
be slim. I thought how, in this dangerous hour, we had completely abandoned our bird protector roles. The cormorants
that were along this shore were now scattered and gull chicks
were hiding from us in the dense vegetation in the middle of
the island. Somehow even the thought of protecting the birds
seemed absurd. In the distance we could see Eastern Egg Rock.
At this hour, Barbara, Tracey, and Debbie were probably counting up their daily bird totals. How trivial that all seemed. Nothing seemed important except to get some help for Roger and
find a way off this barren rock.

But just as birding seemed the least important thing, it
became our salvation. It was about 5:30 p.m., and in the distance, we spotted the *Hardy III* making its regular evening run

The *Hardy III* on its evening cruise to Eastern Egg Rock spotted our frantic signals for help. Photo by Derrick Z. Jackson

from New Harbor to Eastern Egg on its puffin-watching cruise. We stoked up the fire and started wildly waving a blue tarp. Fortunately, another bit of equipment that washed up was Mike's light reflector, which he used for adding fill light during photo shoots. Now Mike used the three-foot reflector to catch the attention of the *Hardy III,* which was at least half a mile away.

Much to our amazement and delight, the boat shifted course and came toward Western Egg Rock. Captain Leonard Duffy was at the helm, and later we learned that his teenage deck hand, Matthew Sampson, had spotted Mike's flashing signal and recognized it as a distress alarm. Captain Duffy approached close enough to see the *Lunda* belly up and people shaking a blue tarp, but he also saw the huge rollers that were breaking on the shore and wisely backed off to a safe distance. The *Hardy* radioed to the coast guard, confirmed that help was on the way, and then resumed her course for Eastern Egg Rock to continue the puffin cruise.

The fog had rolled back in, and it was nearly dark when the coast guard cutter *Wrangell* appeared. We had been on Western Egg Rock about four hours, but the ordeal seemed to go on for days. The coast guard crew promptly sent a Zodiac inflatable boat ashore bearing four capable young men with blankets and a stretcher. All business, they put Roger on the stretcher and carried him to the shore as we trailed along in the near darkness. Relieved at our rescue, I reflected on Ginny Peterson's caution before the trip: "Take care of Roger—he is a national treasure."

Roger later recounted that, while lying on the stretcher looking skyward, he thought he saw a Leach's storm-petrel, and we learned that we were especially fortunate that day because it was uncommon to have a rescue boat available in Boothbay. Roger was admitted into the Saint Andrews Hospital in Boothbay Harbor for a couple of days to recuperate, and the rest of us went back to the puffin house, shaken but happy to be alive. We had left most of the washed-up gear on the island, including the *Lunda,* which was heavily damaged when she came ashore.

A day later, Mike Reny, a local lobsterman from nearby Round Pond, found Roger's camera bag with some of his film floating two miles from Western Egg Rock. Most of the film was in surprisingly good condition in sealed plastic containers. But his Nikon cameras and a four-thousand-dollar telephoto lens were lost or ruined. Mike's films were recovered, but not Judy's recordings. The final cut of "Celebration of Birds" includes some of the salvaged Egg Rock scenes but not our interview.

About a week later, I joined a salvage team headed by Joe Johansen to Western Egg Rock. We found the *Lunda* where we left her, upside down on the south end of the island, bow rail, steering console, and motor damaged beyond repair, but with the hull in fine shape. We rolled her over and towed her back to Hog Island. Twenty-five years later, and rebuilt three times, the *Lunda III* remains our primary boat for Egg Rock runs. For first-time passengers, I usually don't mention the Roger Tory Peterson debacle until we are safely back on the mainland and have taken off our life jackets.

In late fall 1990, I was at a meeting of the International Congress for Bird Preservation in Christchurch, New Zealand, and Roger was the featured speaker. His topic was the importance of bird-watching. I shuddered when he began his comments with an account of our near disaster and hoped that he would not mention my name. He began by saying, "Bird-watchers saved my life," referring to the *Hardy III* puffin watchers. He generously put a great spin on the tale and talked of the "rogue wave" that sent him flailing into the surf. Rather than ending his famed career, the experience rejuvenated him, confirming that at eighty-one he was vital enough to survive the same waters that had claimed his first wife.

Three years later, in response to my fund-raising letter, he sent Project Puffin a personal check for $150, calculating $1.00 for each species that we had seen on our recent birdathon. His typed letter said that he was sorry for not returning to Maine in recent years but that when he did next return, he "probably should keep off small boats."

11

Filling the Ark

Intense hunting of adult puffins and terns for meat and feathers probably eliminated the original seabird population at Eastern Egg Rock in a few short years, but reassembling these lost populations has taken decades. Every species of seabird (except human-wary cormorants) now nests at Egg Rock. These seven avian acres are now home to more than seven thousand nesting seabirds—all tributes to our focused restoration efforts. Serving as seabird stewards for this varied flock requires ongoing creativity, anticipating threats and seeking creative ways to help the birds thrive. Eastern Egg is much like an ark that we have filled with seabirds. We are as responsible today for their well-being as Noah was for his biblical passengers.

Today puffin-seeking tourists can circle Eastern Egg Rock, where more than a hundred pairs of puffins nest in association with about a thousand pairs of terns, including Maine's largest colony of endangered roseate terns. Only three other islands between Massachusetts and Nova Scotia have all three terns nesting, and only Eastern Egg Rock (as of 2014) had at least fifty pairs of all three species.

The *Island Lady,* a seabird-watching boat from Boothbay Harbor, brings puffin watchers to Eastern Egg Rock. In some years more than seven thousand people circle the island to view seabirds aboard the *Island Lady* and *Hardy III;* the companies donate part of their receipts to Project Puffin to help manage the island. Photo by Derrick Z. Jackson

Today, Eastern Egg Rock is also home to laughing gulls. Like terns, these small native gulls were displaced by the larger herring and great black-backed gulls and presently nest in Maine only under the shelter provided by terns. About two thousand laughing gull pairs now nest on the island. They are generally not predatory to puffins or terns the way herring and great black-backed gulls are. But "gulls will be gulls," and these raucous birds sometimes snatch up some tern eggs and chicks as part of their varied diet of insects, fish, shrimp, blueberries, and even small mammals—which they capture on the mainland and sometimes cough up to feed their chicks on Egg Rock.

Their biggest effect, however, is crowding out the smaller terns by their sheer numbers and fertilizing the weeds and shrubs, resulting in a dense tangle of stems that makes the island less suitable for terns, which prefer more open space. For this reason, resident interns now must paint the eggs with vegetable oil to prevent hatching.

Eastern Egg Rock is also an inviting place for more than

two hundred species of migratory land and waterbirds that find sanctuary here. These migrants represent about 25 percent of all North American bird species, along with migratory monarch, red admirals, and angelwing butterflies, as well as the occasional whale, dolphin, seal, and sea turtle.

Other seabirds are also returning to Maine islands in response to management. In June 2009, interns on Matinicus Rock were stunned by the discovery of the first recorded common murre egg (later likely destroyed by a gull). This was the hard-won prize of a social attraction program using dozens of decoys and a playback system at Matinicus Rock. Murres have not nested on Matinicus Rock since 1883, when the species was lost from Maine islands because of hunting for meat and eggs.

And in 2009 a pair of Manx shearwaters were discovered raising a chick in a burrow located near the Matinicus Rock lighthouse, on a section of the island where vigilant interns prevented the nesting of herring and great black-backed gulls. This was the first fledgling-age chick discovered in the United States. The Manx shearwater is a crow-sized cousin of the albatross that usually nests on the British side of the Atlantic. Noted for their extreme longevity, Manx shearwaters may fly as many as five million miles over their half-century lifespan.

The discovery of the first Manx shearwater chick at Matinicus Rock was a long time in coming and points to how long some seabirds take to establish colonies. Of course, I had become very much aware of this characteristic, given how long it took puffins to finally colonize Egg Rock. But here was an independent example of the natural colonization process, how tenuous the first nest can be, and how chance plays out in documenting a new colony. Manx shearwater colonies are especially difficult to discover because they are nocturnal and

Matinicus Rock, Maine's most remote seabird nesting colony, was the only Maine nesting place for puffins when Project Puffin began. Now the puffins are joined by nesting Manx shearwaters and common murres. Photo by Stephen W. Kress

build an underground burrow. The opening of their burrow is a small hole in the sod that is usually hidden in thick vegetation. For these reasons, finding the first breeding birds is difficult, even when there is a field station of keen biologists living nearby and also because first breeding attempts often just "blink out" by chance event without anyone noticing.

At Matinicus Rock, Manx shearwaters started visiting in 1997 and even started digging burrows. An egg was discovered in 2005, but it failed to hatch. But even as the count of the shearwaters reached nineteen birds, no fledged chicks were detected until the 2009 discovery. In that year Project Puffin and Fish and Wildlife Service biologists found a nearly fledged shearwater chick with down between its legs. Now several chicks are banded each year, and these will help to build the colony.

During the fortieth anniversary of Project Puffin in 2013, my interns asked if I envisioned the seabird diversity that we now enjoy when we began transplanting the Newfoundland puffin chicks. I explained that when we first started, we were so focused on puffins that we did not give the other species much thought. But in those eight years of waiting for puffins, we had a lot of time to consider opportunities for concurrent projects with other species such as terns, razorbills, murres, and storm-petrels.

For the Maine coast, this meant a shift to replicating a complete nineteenth-century avian ecosystem. If I had started out trying to restore the whole community, I would probably have had even more people thinking the idea was wacky.

But the length of time it took for this winged menagerie to take shape and take hold is sobering. We started with just 6 chicks in 1973 to build confidence around our techniques, increased the number to 64 in 1974, and increased again to about 100 chicks a year through 1986. When we stopped collecting chicks, we had brought 954 from Newfoundland and fledged 940.

Of all those years, the class of 1977 was the one that made the difference. And this was the group that we almost did not transplant until the last hour. By all counts, fifty-one of the ninety-nine puffin chicks in that year returned to the Maine coast—mostly to Egg Rock. We will probably never know why that year was so successful. Most likely, it was a combination of our methods and ideal ocean conditions (abundant food and absence of extreme storms) that favored high survival of that year's chicks. They were the anchor onto which subsequent transplant groups collected. If we had stopped transplanting puffins after the first four years and quit because we had seen nothing return, I am convinced that we would never have learned the outcome of Project Puffin. Without this perseverance, the idea of restoring seabird colonies would have had a serious setback.

Even though puffins and terns were now nesting at Egg Rock and the prospects for growth were promising, I would occasionally hear from naysayers that perhaps the birds would have returned to the island on their own—even without my efforts. As proof that the bird communities were not changing on their own, I pointed to the great black-backed gulls that continued to dominate nearby at unmanaged Western Egg Rock while the high diversity community of terns and puffins now flourished a few miles away at Eastern Egg Rock. But I was itching for a new seabird challenge and figured the best way to make my case for the success of the methods was to take on another island.

After two years of attempting to get permission to work on Seal Island National Wildlife Refuge, we finally received an approval in 1984 and started a second puffin restoration project about forty miles to the east of Egg Rock. The island had a two-hundred-year-old history as a fishing camp and even had several seasonal wood-frame houses until it was cleared of human occupants and claimed by eminent domain by the federal government so that the US Navy could use it as a bombing target from the 1940s to the early 1960s. Fortunately, for the birds and our restoration goals, Seal Island was transferred to the Fish and Wildlife Service and became part of the Maine Coastal Islands National Wildlife Refuge in 1972.

I returned to Ralph Palmer's *Maine Birds* for a history of Seal Island. And I reread his reference to Arthur Norton's comment that Seal Island was once "the largest breeding site for puffins in mid-coast Maine." That is, before 1887, when puffin hunters killed the last of the original population. Nearly one hundred years later, I was eager to replicate the success we achieved at Eastern Egg Rock at Seal Island. I reasoned that if we could create a second puffin and tern colony, we would certainly demonstrate to all naysayers that the methods for puffin recolonization were valid and that people could actually have a say about which seabirds would frequent the Maine coast.

Yet the obstacles were enormous. First, there were worries about unexploded ordnance from the navy's bombing days. This worry had credence, because a runaway campfire in 1978 ignited the peaty soil on the south end of the island, setting off long-buried munitions. And searches elsewhere found more ordnance buried intact in the peaty soil. Presumably, another fire or a missed step could detonate the aging munitions. Before we received permission to set up the field camp, the navy returned to establish trails so that we could move about the island in safety.

Compared with Egg Rock, everything at Seal Island was on a much grander scale. First, Seal Island was nearly three times as far from the mainland as Egg Rock. This distance would add greatly to the costs of operating a field camp on the island. And then there were the gulls. The island had become one of Maine's largest gull colonies, with at least 2,600 pairs of herring and great black-backed gulls. I was convinced that the presence of the gulls had prevented puffins from naturally re-colonizing the island and that their continued presence would prevent us from successfully reintroducing puffins and terns.

In a cooperative program with Fish and Wildlife and the Canadian Wildlife Service we moved forward with plans to revive this sleeping giant of Maine seabird nesting islands. The Fish and Wildlife Service took the lead on gull control and for several years starting in 1984 used the avicide DRC-1339 to kill the nesting gulls. As at Egg Rock, many succumbed, but the survivors learned to be wary. With time, it became apparent that the most effective way of scaring gulls off the north end of the island, where we planned to rear and release puffins, was to hire interns with marksmanship skills. Gulls quickly learned it was dangerous to be anywhere near the northern end of the island and kept a safe distance. Not surprisingly, firearms (the primary tool of the market hunters) were once again our best tool for managing gulls. Except now we were counting on the gulls' intelligence to avoid habitat where they were at risk of a twenty-two-caliber rifle.

The gulls were so ubiquitous at first that we started to attract terns using decoys and sound recordings behind the research cabin, a twelve-by-twelve-foot plywood shelter modeled after the Egg Rock Hilton. As gulls began to become more wary and loosened their grip on the island, we moved the tern attraction equipment consisting of fifty decoys and a sound system to the vicinity of the puffin burrows, because I wanted the puffins to benefit from the feisty nesting terns. Given the difficulty of displacing gulls and the reality that terns had not bred at the island for thirty-five years (leaving no adults with a memory of nesting here), it was not too surprising that it took five years before the first of sixteen pairs of common and Arctic terns nested among the decoys in 1989. Twenty-five years later, the tern colony has grown to be the largest in Maine with about 2,200 pairs (about 60 percent common and 40 percent Arctic).

David Nettleship agreed to renew our puffin-collecting trips to Great Island, and these resumed in 1984 with the condition that we conduct diet studies to measure the growth of chicks on various kinds of foods. These studies were to be supervised by Canadian seabird researcher Tony Diamond. In all, we moved 950 chicks through 1989, fledging 912 from sod puffin condos.

I was convinced that puffins would eventually return to Seal Island as they did to Eastern Egg Rock, but the big unknown was what effect Matinicus Rock, only six miles away, would have on this project. The initial returns of puffins only added to my worries. Like the first returns to Egg Rock, the first returning puffins from 1984 visited Seal Island but chose to nest at Matinicus Rock. The project got off to a poor start, with few returning birds from the transplant groups fledged between 1984 and 1986. Of the 347 puffin chicks fledged in these years, only 20 returned. Fortunately, high survival of the 1987–1989 transplant groups provided the critical mass necessary to found the new colony.

The most impressive showing was from our class of 1988. An amazing 99 (53 percent) of the 188 birds released in 1988, and an impressive 57 of the 190 fledged in 1989, were eventually resighted. This result reaffirmed my belief in the importance of multiple years of releases. It also hardened my suspicions that the frequent handling of the chicks in 1984, 1985, and 1986 (to conduct the feeding studies) had somehow affected the chances of returning. Perhaps the handling had damaged the waterproofing of the growing chicks or stressed them too much. The message was clear: we needed to minimize handling as we had done in the successful years at Egg Rock.

Then in 1992, like a proud father, I announced again the birth of a new puffin colony—there were just seven pairs, but each was a miracle of patience and perseverance. Again, there was an eight-year wait from moving the first puffin chicks to

Seal Island is home to more than five hundred puffin pairs. Observation blinds dot its perimeter. Here, puffin web cams beam real-time video to the Internet. Photo by Stephen W. Kress; inset by Derrick Z. Jackson

seeing puffins carrying fish to a new generation of puffins. Indeed, Seal Island has proved to be a much larger success than Egg Rock. Today more than five hundred breeding puffin pairs produce hundreds of young each year.

Likewise, the decades of management on Matinicus Rock, where Audubon-paid wardens saved the last surviving puffin pair more than a century ago, has grown to about 350 breeding pairs. In 2011, interns reached into these puffin burrows to band a record two-hundred-plus puffin chicks.

The resurgence of Maine puffins has led to other new puffin and tern colonies. Notable is Petit Manan Island near Bar

Harbor, part of the Maine Coastal Islands National Wildlife Refuge. Petit Manan hosted fifteen hundred nesting terns around 1970. But when the coast guard automated the lighthouse in 1972 and left the island to the birds, black-backed and herring gulls invaded. By the early 1980s, there were no nesting terns left. In mid-May 1984, the US Fish and Wildlife Service, with the help of the College of the Atlantic in Bar Harbor, launched a program of gull control and tern restoration similar to what we did on Egg Rock, starting with the same avicide that Frank Gramlich used a decade earlier at Egg Rock—poisoned bread cubes placed in nests.

Leading the College of the Atlantic's effort in the project was none other than the man who helped me break the ice for my puffin restoration idea with Nettleship: Bill Drury. He had left Mass Audubon in 1976 to move to Bar Harbor to become a professor at College of the Atlantic. More than a decade removed from defending me, he had to break yet more ice to bring the terns back. There was enough local outcry about poisoning gulls in the nearest coastal town of Milbridge that Drury attended a community meeting to explain what they were doing. The *Bangor Daily News* of May 1, 1984, reported how Drury reminded the audience how terns had disappeared from Massachusetts and how gull populations in Maine had exploded twentyfold since 1900, flourishing on garbage, sewage, agricultural waste, and waste from the fishing industry.

The audience was sufficiently convinced, and the poisoning commenced a couple of weeks later. Forty-eight hours later, researchers hauled off the carcasses of approximately 670 herring and great black-backed gulls from Petit Manan Island. The terns seized the opportunity to reclaim the island and were back on territory within a week of the gull treatment.

Although coastal dumps have been closed, gulls still find abundant food at "sanitary" landfills, especially in winter, which is a season that historically was most difficult for gulls and probably was a natural check on their numbers. Abundant garbage throughout the year has made life much easier for these adaptive birds.

By the end of June 1984, Drury recounted in his memoir that there were 855 tern nests at Petit Manan. "Some argued against the killing of gulls on humanitarian grounds," Drury wrote. "I agree: killing is not a pretty or enjoyable activity. . . . But gulls eating living tern chicks is not a pretty sight. . . . I think the philosophical question of killing one species to favor another was made and accepted by those early agriculturalists who pulled up plants that inhibited the growth of their crops—they weeded the garden."[1]

In the case of Petit Manan, that garden was so well "weeded" that puffins, of which there was no prior record of breeding, began landing on the island in 1984, likely because the presence of terns suggested that it would be a safe zone from gulls. Although there were few rock crevices for natural nesting burrows, a few pairs began nesting in 1986. The first banded native puffin chicks from Petit Manan began returning in 1994 and began breeding in 1996. The colony increased to seventy-three breeding pairs in 2014, with high counts of almost two hundred puffins. Six of these pairs nested in artificial burrows built by the refuge managers; most of the others either burrow in the shallow soil or prefer shallow rock crevices. Even though there are islands nearby with better nesting habitat, the puffins chose Petit Manan, apparently because there were fewer gulls and more terns. The presence of the resident island stewards was clearly not a deterrent for either the puffins or terns. Those seventy-three pairs helped push Maine over the one-thousand-puffin-pair mark on four islands, ten times the number that resided solely on Matinicus Rock in the early 1970s.

Bill Drury died in 1992, the year that puffins nested for the first time on Seal Island, one of his favorite islands. Happily, he saw the response of puffins and terns to applied management on his beloved Maine islands. None of the work described here would have been possible without his vision.

Then in 2006, John Anderson, Drury's successor at College of the Atlantic, discovered a single pair of puffins nesting at Great Duck Island, located roughly midway between Seal Island and Petit Manan Island. This nesting provides further evidence of the growth and expansion of the Maine puffin population. Oddly, the single pair nests just a few yards from an especially aggressive pair of nesting great black-backed gulls in a part of the island that has seen a recent increase to twenty-six pairs of black-backs. Other puffins are now associating with these pioneers, who have now managed to hatch their egg every year from 2006 to 2012. Time will tell how well this tenuous arrangement will fare, but this nesting demonstrates surprising variability in puffin behavior. Although most birds may be shy and retiring around gulls, occasionally there will be an especially brave pair of "pioneers" who will nest alone and attract followers.

The restoration of the puffin in Maine was superficially less important than the reestablishment of bald eagles in the contiguous forty-eight states, where the species was down to its last 487 pairs in the 1960s, or peregrine falcons, which were wiped out in the eastern United States. In the case of peregrines and California condors, the species required captive breeding programs, which add another big level of human engagement. Though the approaches taken were far more expensive than anything I have tried with Project Puffin, these recoveries were more straightforward to orchestrate.

Removing DDT had almost miraculous effects, as we've seen over the past half-century. Bald eagles, aided in many states by translocation of chicks and eggs, now nest again in all forty-eight contiguous states. Peregrines now soar from national forest cliffs and urban skyscrapers. Once completely lost from the eastern United States, peregrines are back, with more than a thousand nesting pairs by 2012. Although the original population nested mainly in natural habitats, now most are city dwellers, occupying urban canyons and bridges.

Puffins, in contrast, are much less adaptable to human-dominated landscapes. Their specialized diet and nesting habitats dictate that they do not benefit from either garbage or coastal development. Life in the middle of the food chain is a tough place to live. Puffins are vulnerable from below by shifts in their food and ecosystem level changes, and they are hammered from above by burgeoning predator populations such as gulls, eagles, and falcons.

In the years following the restoration of the Seal Island puffin colony, it became increasingly clear that terns were in need of active management on the Maine coast and that it would not be sufficient to manage them just at a few colonies. In an ambitious growth period for Project Puffin (which is now called the Audubon Seabird Restoration Program to denote its broader interests), we set out to manage terns along the entire middle and southern Maine coasts. Like a recipe, we first displaced the gulls, then used social attraction, and followed up by setting up a field camp with summer interns to protect the

Island / site	Year started	Nesting waterbird species	Notable species
Eastern Egg Rock (MDIFW)	1973	11	Atlantic puffins, roseate and common terns
Matinicus Rock (USFWS)	1978	14	Atlantic puffins, razorbills, Arctic terns, Manx shearwater
Seal Island National Wildlife Refuge (USFWS)	1984	13	Atlantic puffins, Arctic and common terns, great cormorants
Stratton and Bluff Islands (NAS)	1986	19	roseate, least, and common terns, egrets, ibis, water-fowl, shorebirds
Jenny Island (MDIFW)	1991	4	common terns, roseate terns
Pond Island National Wildlife Refuge (USFWS)	1996	4	common terns, roseate terns
Outer Green Island (MDIFW)	2002	4	common terns, roseate terns
Hog Island Audubon Camp (NAS)	1936	—	—
Project Puffin Summer Base	1986	—	—
Project Puffin Visitor Center	2006	—	—

restored colonies. Between 1986 and 2002, we established field camps at Stratton Island, Jenny Island, Pond Island, and Outer Green Island. These islands, combined with Eastern Egg Rock, Matinicus Rock, and Seal Island, were now providing habitat for about 80 percent of Maine terns and 95 percent of the state's puffins.

During these years, my own "ark" took on some additional passengers. Evie and I married in May 1983. The first leg of our honeymoon took us to Wooden Ball Island, in outer Penobscot Bay, where we occupied a rotting fisherman's shack with a floor that was so far gone that I swept up earthworms while attempting to make the place habitable. We were there in part to test the idea of using puffin decoys to attract puffins, hoping that decoys alone might result in a new colony just four miles from Matinicus Rock. Our time on Wooden Ball was followed by another unlikely honeymoon destination—Iceland. Here we camped among snowdrifts, swam in hot springs in deep lava grottos, and reveled at immersion into the center of the puffin's greatest empire.

Our son Nathan was born in 1985 and his brother Ben in 1986. We carried them as little bundles to Eastern Egg Rock and Matinicus Rock, spending summers on Hog Island, winters in Ithaca, New York, exposing them to the Maine coast and other adventures, including three trips to the Galápagos Islands. When still a pre-toddler, Nathan rolled into an Egg Rock "spooge pool" of seabird guano one frigid May while Evie and I were preparing tern habitat. Dried off, he was good to go for the next adventure, which was never far away. Later, he took his first steps aboard the Audubon camp's *Puffin III*. Now he is making a career of coaching crew and teaching history, passions perhaps rooted in many bouncing boat trips in the

Lunda across Muscongus Bay. Likewise, Ben has an emerging science bent, focused on neurology and a medical career. I like to believe that he began thinking about biology while hearing me go on about birds and the importance of interconnections and the wonder of life. Evie and I began to drift apart and divorced in 1997. Five years passed before I decided to try marriage again.

Elissa Wolfson and I first met at Hog Island in July 1989; she had received a scholarship to attend a Hog Island session. Concerned about the project's gull management techniques, she asked pointed questions about gull control following a lecture. About ten years later, our paths crossed again in Ithaca. She was then editor at Cornell Plantations, the university's botanical garden. She interviewed me about attracting backyard birds for an article she was working on. A year passed, and I found that she had signed up for my spring field ornithology course. I asked about the article, only to learn that her pocket tape recorder had malfunctioned and the interview had been lost. After a second interview, I offered to provide photos to accompany the article. That led to our first date in spring 2000. We tumbled in love and then into marriage in 2002. Rabbi Steve Shaw provided nuptial blessings at our rural Ithaca home. No surprise, as a loyal Hog Island alum and puffin enthusiast, he found ways to weave puffins into the ceremony.

We began our honeymoon where we had first met, with a quick trip to Hog Island, where I was teaching a birding program. Part two of the honeymoon came later, when we visited Bermuda (a more conventional honeymoon destination than Wooden Ball Island), where we also volunteered to assist naturalist David Wingate on Nonsuch Island. There we spent most of our time poking around in cahow burrows and helping Wingate remove poisonous toads from Nonsuch. Five years later,

Elissa and I became the proud parents of a daughter, Liliana
Pearl Kress. She, too, migrates with us to the Maine puffin islands each summer: first as a little bundle, then as a tiny toddler, and now as a child of nature with an eye for birds and a
love for the ocean that runs neck and neck with her love for
princesses.

12

Project Puffin Goes Global

I n 1973, when I was moving the first puffin chicks, I could never have guessed that my dream of restoring a puffin colony would morph into a lifelong quest. But even more remarkable are the various applications that have originated from our Egg Rock experiments. Methods such as social attraction and translocation of seabird chicks are now standard practice in the toolbox of seabird managers worldwide.

These methods originally helped me bring seabirds back to Maine islands after the seabird massacres associated with the Victorian era feather trade; since then, the methods have benefited birds jeopardized by invasive mammals, egging, fisheries conflicts, oil spills, and flooding due to climate change. In every case, humans have either directly assaulted the seabirds or changed their world through some indirect action. Even Torishima Volcano, which threatens the short-tailed albatross, would likely not be an extinction threat without the avarice of humans who killed millions of the birds for feathers in the late 1800s, making them vulnerable to volcanic eruption.

Some seabird restoration projects also help other species by restoring habitat and ecological integrity. For example, res-

imitation of burrowing petrels in New Zealand is also creating burrows for rare reptiles such as the New Zealand tuatara.[1] By restoring lost seabird colonies, entire ecosystems can benefit by importing marine nutrients through seabird guano to islands to fertilize indigenous plant communities. Seabird guano can transport more than a hundred times as much nitrogen and four hundred times as much phosphorus as rainwater.[2] Excessive amounts of seabird guano, especially in arid climates where it accumulates on land (such as at cormorant and booby colonies) can devastate vegetation, but moderate amounts, delivered by burrowing seabirds that also aerate the soil, can result in lush vegetation that reduces erosion throughout the year. The benefits to marine ecosystems surrounding seabird islands are difficult to quantify, but it is likely that nutrients washing off islands also fertilize marine algae that are home to a myriad of ocean life.

The global effect of our Maine seabird program was not planned. There was never a strategic plan written by a high-priced consultant. It was completely organic: a good idea that came along in its own time, recognizing that human impact is so universal that it was no longer appropriate for people always to simply stand by and let nature take its course. Too often that ended in extinction as populations and ranges collapsed from local extirpations. Most of the restoration projects that are helping seabirds today are possible because of individuals who have heard about our success with puffins and decided to take action to adapt the method to other species.

The list of applications for seabird restoration continues to amaze me, and opportunities to apply the methods will certainly increase as a way to counter some of humanity's many negative impacts on wildlife. Behind every restoration are in-

novative and caring biologists who are doing their best to enhance chances for survival for rare species and enhance biodiversity. These are true heroes of our time, often risking their
own safety to undo mistakes of others.

In a 2012 review for the *Journal of Wildlife Management,*
conservation biologist Holly Jones and I detailed a few case
histories and conducted a world review of seabird restoration
projects that have used social attraction (decoys and sound
recordings) and translocation of seabird chicks. We found that
128 seabird restoration projects had been tried around the
world to benefit forty-seven seabird species in a hundred lo-
calities in fourteen countries. In analyzing some of the better-
documented projects, we found fifty-nine that used decoys
and audio recordings to lure seabirds to new nesting habitat;
fourteen projects that used only acoustic playback; ten that re-
lied solely on chick translocation; and eight that depended on
decoys alone. A combination of transplanting chicks and audio
recordings was used in nine more projects, and three others
combined translocation of chicks with decoys.[3]

Eighty-eight carefully documented projects had been
under way long enough to measure success. Of these, we found
that fifty-five had been successful in restoring a nesting colony,
and forty-one of these employed some combination of chick
translocations, decoys, and sound recordings.[4]

Project Puffin's international projects started in 1986 when I
was thumbing through the mail late one evening in my office
at Cornell's Laboratory of Ornithology. That sounds like a comfortable location for an office, but I was actually in the back
room of a little, one-story, fiber-walled building that had been
constructed for studying barn owl flight. After owl researcher
Roger Payne left Cornell to begin his career studying whales,

the tiny building was repurposed into a few very small offices. The building was forgotten by the university, and I was given permission to set up an office in 1975. Rent- and utility-free, Owl House was the first home for Project Puffin until we moved to the grand Imogene Powers Johnson Laboratory of Ornithology in 2003.

The notice that caught my eye was from Montres Rolex S.A., the famous watch company. Rolex prides itself on a history of innovation and was announcing a worldwide competition for the 1987 Rolex Awards for Enterprise. The notice explained that Rolex was searching for five enterprising people who were early on the path of novel careers in applied science, exploration, and the environment. The highest prizes would go to five laureates. Each would receive a grant for 50,000 Swiss francs (about $34,000), a gold Rolex chronometer (watch), and a trip to Geneva to attend the award festivities. I had already become accustomed to taking long shots for grants to support the puffin program, and by 1986 I was buoyed by the successful restoration of puffins and terns at Egg Rock. I concluded that I might have a chance because my restoration methods of social attraction and chick translocation were innovative notions that I believed could benefit endangered seabirds—an idea that I used to justify the work with puffins from the outset of the project.

I submitted the application and explained how I would use the prize to help the endangered Galápagos (formerly dark-rumped) petrel, an endemic burrow-nesting seabird that was facing extinction because introduced rats were attacking all four of the remaining colonies. The future for the petrel was ominous: the population of these long-winged seabirds was declining at the alarming rate of 30 percent per year. If that rate of decline continued, the species could be extinct within the next ten to fifteen years.[5] I also proposed using decoys and translocation of short-tailed albatross chicks to a nonvolcanic island.

Months passed without any word from Rolex, and I had nearly forgotten about the proposal when I received a telegram to arrange an interview with a Rolex representative who wanted to fly from Geneva to meet me. We agreed to rendezvous at LaGuardia Airport in New York, a location from which I had other travel plans. The interview apparently went well, because soon thereafter I received another Rolex telegram informing me that I was selected as one of the five grand prize "Lauriat" winners! The winners were a diverse group: Jacques Luc Autran, a French humanitarian who outfitted a trawler to provide medical supplies to islanders of the Maldives; Pierre Morvan, an amateur French insect taxonomist who became the world authority on Himalayan ground beetles; Nancy Lee Nash, a Hong Kong–based author who bridged the gap between Buddhist literature and nature protection; and Johan Gjefsen Reinhard, an anthropologist well known for exploring extreme high elevations to learn about ancient cultures.

Soon after the announcement, Rolex made arrangements to fly Evie, our one-year-old son Nathan, and me first class on Swiss Air to Geneva. My proud parents accompanied us on the trip.

With dedicated funds for my first international project, I reached out to Felipe and Tina Cruz, Galápagos petrel researchers who helped to open doors with the Charles Darwin Research Station and Galápagos National Park. I had been to the Galápagos Islands about a dozen times previously, leading natural history tours for Holbrook Travel. Those tours had given me some happy adventures and grounding in Galápagos natural history, but they also provided experience with the extreme climate and rigors associated with the islands.

On one memorable adventure, we had made a tortuous climb for the better part of a hot day over jagged lava and cactus up to the highlands of Volcán Alcedo in search of giant tortoises. After draining our canteens about halfway, we decided to top off our personal water reserves with water from a communal water jug that was being hauled up the mountain by a weary porter. But it turned out that the porter had lugged a jug of gasoline-polluted water rather than the fresh water we expected. Now all the water was gasoline-tainted. The expedition was immediately jeopardized, and the porter (who moments earlier was a hero for carrying the heavy water jug) evaporated into the wilderness.

Fortunately, one of our group had packed an orange, which we divided among the group. We each discovered that, by holding a sweet section in the mouth, we were able to keep moving and eventually reach the wet zone, where we could gather rainwater off our tents. It was daunting to consider doing research in such extreme habitats, where even the basics of safe water could not be assumed, but the idea of living in the Galápagos without the responsibility of a crowd of tourists was especially exciting.

By May 1988, I was on my way to Santa Cruz Island in the Galápagos with my friend and colleague Richard Podolsky. We packed our suitcases with outdoor speakers, cassette recorders, and stacks of solar panels. We also brought boxes of prerecorded endless-loop cassette recordings of Galápagos petrel calls created by the Lab of Ornithology. The goal of the project was to entice these rare, nocturnal birds to nest in protected artificial burrows by broadcasting petrel recordings. Our successful efforts to establish colonies of the distantly related Leach's storm-petrel in Maine and the Laysan albatross in Hawaii (both subjects of Podolsky's doctoral dissertation) gave us hope that we could apply this technique to the distantly related Galápagos petrel. We hoped that the broadcast sounds would attract the petrels to the artificial burrows within a secure, fenced zone. The predator fence would keep out roaming dogs, cats, and pigs, and rats could also be more easily controlled within the fenced area. In this way, the colony would survive as a secure island within a sea of introduced predators.

Our part of the project was to determine which recordings were most attractive to the petrel and whether they would in fact nest in artificial burrows. The study site was at Media Luna, a long-extinct volcanic caldera in the highlands of Santa Cruz Island. It was here that Michael Harris had documented the precipitous decline of the petrels and the threat from black rats that whalers had accidentally introduced to most of the Galápagos archipelago.

The national park service provided lodging in its field camp high on Santa Cruz Island, where we were given the use of a tiny casita (hut) notable for the dark mold that grew on its walls and everything else that stood still for a few hours. We shared the hut with several park employees assigned to battle *Cinchona officinalis.*

A tree grown for its reputation as a one-plant pharmacy, cinchona is used for treating digestive disorders, circulation issues, hemorrhoids, influenza, and cancer. It is best known for treating malaria because the bark is the source for quinine. Yet despite its benefits, in the Santa Cruz highlands, cinchona was invasive plant enemy number one, and the Galápagos National Park was making a valiant effort to keep it out of the Media Luna region, where it could overrun important native plants such as miconia, an endemic shrub found only in the highlands of Santa Cruz and San Cristóbal Islands.

The park employees were charged with hand-pulling cinchona saplings before they could establish deep roots. They laid these on top of other plants so that the roots would dry—which was not always effective in the usually soaked climate. Cinchona trees that were too large for pulling were hacked to the ground, but they would soon sprout and required repeated trimming to reduce the production of more seed. The more fortunate cinchona pullers rode horses, but most were on foot, pulling for long hours, each equipped only with a machete to attack the larger plants and a large plastic bag to wear as a raincoat.

They were a fearsome sight with machetes flailing, dressed in black plastic, but at night the cinchona pullers were amiable companions in the casita. Despite their primitive approach and seemingly endless supply of cinchona, it was heartening to see their results: there was a distinct wall of cinchona just on the edge of the national park.

It was little wonder that the walls of the casita grew mold. On most days Media Luna was in the clouds, windy and wet, especially when we were soaked by the *garua*. Somehow this Spanish word for drizzle captured the bone-chilling wetness better than any English equivalent. On most days, the garua was so thick it was nearly edible. Raincoats and pants were necessary daily attire. Oddly, on such days we felt very much at home here because of similar weather on the Maine coast.

Occasionally, the weather would lift, exposing glorious views to the sea some ten miles away. Without wind, the silence was stunning—just the occasional chirp of one of several species of Darwin's finches. Most notably, there were no motor sounds. No planes, boats, or vehicles to remind us of human presence. Yet the effects of humanity were present everywhere in the shape of invasive cinchona and guava that escaped from nearby farms along with rats that lurked in the thick and tangled vegetation.

This project had greater importance beyond any help that we might offer the Galápagos petrel. These petrels belong to the genus *Pterodroma*, which comprises thirty-three species and has the unfortunate distinction of being one of the most endangered genera of birds in the world. The group is especially vulnerable because the young show a very high tendency to return as adults to breed at their natal colony. This habit, called philopatry, has led to many isolated populations on remote islands. Before the human introduction of mammals to remote islands, philopatry made ecological sense because young birds could simply return to the place where they hatched and be likely to find a safe nesting island. But the behavior backfires when invasive mammals arrive. Even the most remote corners of an island are not safe from four-legged invaders. Some petrels are making last stands on rugged mountaintops and tiny islets because people, pigs, dogs, and rats have eliminated them from the more hospitable island habitats in places such as Hawaii, Dominican Republic, Haiti, and New Zealand, where they lived in vast numbers before people and their commensal mammals arrived.

Our first objective was to compare the attractiveness of the different recordings with variable numbers of calling petrels. This information would help us know which recordings to play for attracting birds into artificial burrows. Richard and two Ecuadorian assistants played the recordings over the course of the 1988 summer. Solar-powered cassette players broadcast the recordings from under mist nets strung on an elaborate system of pulleys and ropes fifty feet off the ground. Our intent was to see how many petrels we could capture with different recordings that were randomly played under the nets. While they were doing this work, I headed back to Maine to continue the work with puffins and my teaching on Hog Island.

The Santa Cruz team captured more short-eared owls than petrels in the first days of the project, but the petrel activity increased as the season proceeded, and by mid-August 282 of the endangered birds had been netted and banded. Plumage characteristics informed us that nearly all of the captured birds were too young to breed, which was heartening to know because these were the birds most likely to be attracted to artificial burrows. Of seven tapes played, the most attractive was a double-intensity recording of normal colony sounds recorded to simulate a larger colony. This discovery supported the idea that young petrels prefer to frequent areas where there are larger numbers of birds. Apparently, once Galápagos petrels reach breeding age, they return directly to their nesting burrows after spending several days feeding at sea. Here, they defend their burrow from intruders, incubate an egg, and later brood and feed their single chick. At their burrows, the petrels were notably quiet. Most of the calling was in the air from young birds looking for a social scene and prospecting for a nesting burrow.

In addition to the capture experiment, we hand dug eighty artificial burrows into the volcanic soil of Media Luna—one of several extinct cinder cones a short walk from the casita. Here we set up sound systems near the burrows to broadcast petrel calls. Digging these burrows was reminiscent of excavating the puffin burrows at Egg Rock, except that the volcanic soil permitted rain to pass through and there were no issues with flooding. We set toothpicks in the entrances of the finished burrows and spread soft ash to capture tracks of landing petrels. Five days after the start of the burrow experiment an adult petrel was discovered peering out of one of the hand-dug burrows. By the end of the summer, over 61 percent of the artificial burrows showed signs of petrel activity, and we found sixteen birds in the burrows. We were thrilled that the birds were so responsive.

In the second year of the project, Richard and the two Ecuadorian assistants repeated the netting experiment. Over the two years, they netted six hundred petrels during 868 hours of playback and built 160 artificial burrows. Petrels continued to visit the artificial burrows, and by the end of the 1989 season, 71 percent of the 160 burrows showed signs of activity, including one with an egg. In 1990, "Seabird Sue" Schubel and Jerry Skinner, an instructor from Hog Island, continued the project with Richard and me helping to set up in May. Our two Ecuadorian assistants, Manuel "Manolo" Fajardo and Washington "Wacho" Arevalo, were from the University of Guayaquil and had marine biology degrees. They explained that the only job opportunities for them to pursue marine biology would be to take aquaculture positions at Ecuadorian shrimp farms. We hoped to offer them more experiences in seabird conservation so that they would have a better chance for conservation positions in Ecuador and invited them to Maine for the summer of 1990 as our first international Project Puffin interns.

By summer's end, four pairs of Galápagos petrels nested in the artificial burrows and many other burrows showed signs of activity. Rats, however, killed three of the first chicks, a sober reminder that, although the attraction project was successful, the ultimate benefit would require ongoing predator management.

The petrel fence was never built, and the social attraction equipment that we brought to the island is probably tucked away in storage boxes in a corner of a shed (or tossed out, which may be appropriate considering that twenty-five years have passed since Richard and I dug those first petrel burrows). Fortunately, the Galápagos National Park has continued ongoing use of rat and mouse poison eight or nine months each year near existing petrel nests at Media Luna. This practice is more important than ever, because the larger Norway rat has moved into the area, adding a new threat along with well-established black rat and house mouse populations.

In many ways, this control is similar to the tedious work of the cinchona pullers. It needs to be repeated every year by experienced people. Control, even at this minimal level, depends on annual funding and local knowledge about where the petrels nest in the dense vegetation. Such efforts are vulnerable to changing personalities and changing policies and priorities. At best, it is a stopgap approach until better methods are in place for a more effective way to clear predators from large islands such as Santa Cruz.

There is hope, because increasingly large islands are being cleared of rats through use of rodenticides distributed by helicopters and on-the-ground treatments. In 2012, such a project was conducted at Pinzón Island, a 1.9-square-mile island in the Galápagos. With a cost of about three million dollars, it is not the most expensive project of its type to date or the largest, but working in the Galápagos is especially challenging because of the diversity of endemic and endangered species that must be protected from exposure to the toxic rodenticides that are used for such large-scale eradication projects.

Yet these methods are being considered for Floreana Island, a sixty-seven-square-mile island that supports one of the four remaining populations of Galápagos petrel. The challenges here will not only be the much larger size of the island and the enormous cost and protection of native wildlife but also the greater risk of rodent reinvasion, because Floreana has a resident human population of about a hundred and is a popular tourist island. If successful, this will be the largest project of its type in the world. The record at present is forty-three-square-mile Campbell Island, New Zealand, where a helicopter eradication plan successfully removed Norway rats that were threatening albatross, petrels, and penguins.[6]

Santa Cruz Island is an even more unlikely place for a helicopter-type rodent eradication program, because it has 381 square miles of rugged habitat with a human population of twelve thousand, many of whom live on farms surrounding the petrel habitat. The idea of removing rodents from Santa Cruz Island is thus a long way off. Considering these obstacles and enormous costs, starting new colonies on predator-free islands makes a lot of sense. Combining active restoration methods such as social attraction and translocation of chicks greatly increases the likelihood of establishment of colonies. Although the installation and maintenance of social attraction equipment costs thousands of dollars, the cost is minor compared to multimillion-dollar helicopter-type eradications, and these methods can help improve the chances that missing species will recolonize their former habitat following an eradication.[7]

Rats, mice, feral cats, and pigs still roam Media Luna, threatening every petrel chick. Gaps in this precarious annual management could mean the end for the petrel. A more ambitious restoration project will take a passionate champion to move a new idea forward. That leader will need patience, persistence, and payroll to remove the predators or start new colonies on predator-free islands. Until then, the Galápagos petrel, like many other beleaguered seabirds, remains at great risk.

The cahow (Bermuda petrel) is a close relative of the Galápagos petrel—another of the endangered *Pterodroma*. Although the cahow population is much smaller than the Galápagos petrel, numbering only about one hundred pairs, it has the advantage of strong advocates positioned within government, a recovery plan, and funding to implement restoration and protection. As is usually the case, there is generally one individual with the passion and dedication to bring about change. The story of saving the cahow precedes Project Puffin. This is a bold rescue story with inspiring heroes.

The cahow is native only to Bermuda. Fossil remains suggest that the species numbered half a million before the arrival of the Spanish in the 1500s. Its name comes from its eerie, moaning cry, which resembles a haunting chant. The cry is so piercing that Spanish sailors at first were too intimidated to land on Bermuda, thinking the island inhabited by demon spirits. The spell cast by the spirits did not last long: the gray-and-white seabird was quickly discovered to be tasty food for starving colonists. Rather than demons, the petrels, the colonists came to believe, were gifts from God: as the colonists huddled cold and exposed around fires, the petrels would sometimes fly into the flames as if they were being delivered from above, literally cooking themselves to the colonists' delight. This observation led to more fires, expressly to attract and capture the birds for food. The subsequent introduction of pigs, dogs, and rats nearly finished off the petrels.

The cahow was thought to have been extinct since the 1600s until a mysterious petrel was found dead under the reflective lights of a Bermuda lighthouse in 1935. That singular report caught the attention of a young Bermudan naturalist, David Wingate, who enthusiastically accepted an invitation in

1951 to join a search for the nearly mythical bird. The curator of birds at the American Museum of Natural History and the head of the Bermuda Aquarium invited the lanky teenager to join a search for the cahow. Much to their delight, they found a total of eighteen pairs on four tiny islets in Castle Harbour. These surviving cahows nested in cavities so inhospitable to human feet that Australian petrel expert Nicholas Carlile described them as "the most horrific place. It's like landing on knife blades and scissors. If you fell you would impale yourself."[8] Although the cahow once nested throughout Bermuda from the highest elevations down to the sea, the only viable places safe from the ravaging pigs and rats were these inhospitable islets that excluded four-legged predators.

But for Wingate, these islets were the last vestige of wilderness. Because of their isolated location, they gave the cahow the protection that they needed, and Wingate became committed to nurturing this relic population. He maintained his interest in birds, graduated from Cornell, and became Bermuda's conservation officer in 1966. He then single-handedly took over the restoration of fifteen-acre Nonsuch Island with the vision to bring back a living museum of a completely restored Bermuda plant community. He hoped that, if Nonsuch were saved, the cahow would eventually relocate there and establish a thriving colony.

To achieve this bold plan, he made nests, nursed sick birds, shot predators, and battled developers. He also ripped out everything exotic and drove out invasive rats and poisonous toads and planted ten thousand indigenous trees and shrubs. The goal: to return to the cahows a paradise whose only four-legged creature was the critically endangered Bermuda skink, yes, a skink.

Wingate's mission and his success were already well known in bird circles when I first traveled to meet him in 1975, just two years after I began restoring puffins in Maine. I was interested in seeing his artificial burrows, given the issues that I was having with my ceramic and sod burrows. At the time, he was experimenting with artificial concrete burrows that would be attractive and safe for the cahow, working out the details of the customized entrance holes that would permit the cahow to enter the burrow but exclude the slightly larger white-tailed tropicbirds.

I arrived in Bermuda having in mind some kind of burrow using PVC tubing, but David warned me not to use PVC, because the floor of the burrow would collect moisture and soak the birds. He encouraged me to develop a burrow design with a porous floor to create better drainage and so that the puffins could dig as they naturally would. This would reduce risk of the puffins becoming too damp. It was a very practical discussion. I found this notion of using the peaty soil very appropriate, because it would permit drainage, and we had an abundance of it. The discussion was also timely because we had already seen issues with damaged feather development in guillemots and puffins from the ceramic burrows that we used on Egg Rock the previous summer.

The idea of using soil-bottom burrows also resonated because after the debacle of raising guillemots in the Binnacle, I too could see the value of replicating natural burrows as much as possible. Even then, David had some cement burrows on Nonsuch Island. We talked about the idea of moving chicks into them, but he wanted to see if they would find them on their own.

Now, forty years since we first discussed seabird restoration, David remains passionate about cahows and is widely recognized as being in the right place and doing the right thing to bring back the cahow from the brink of extinction.

Unfortunately, my first trip to Bermuda ended prematurely. While riding my rented motorbike over the twisty Bermuda roads, the bike hit a patch of loose gravel, and I went flying over the handlebars. Catching myself with hands down to break the tumble, I broke my left wrist and ended up in the hospital. The trip was cut short, but the conversations with David helped guide the success of our puffin burrow construction.

Wingate's artificial burrows and wardenlike vigilance on the islets helped cahows increase to around thirty pairs by 1980 and forty pairs in 1990. But they were still under sixty pairs by 2000. By then, concern was very real about the fragility of the rocky islets—not just in the face of normal hurricanes but also in the long-term forecasts of more intense storms and flooding and erosion from the rising seas associated with climate change.

Fortunately, Wingate's forty years of restoration made Nonsuch Island ready to try a chick translocation project. In 2002, my honeymoon plans with Elissa brought me back to Bermuda, where we had the opportunity to visit with David and his successor, Jeremy Madeiros. They were focused on plans to move some of the cahow chicks off the eroding islets and onto Nonsuch Island.

The plan made obvious sense. The original, tiny islets that had provided sanctuary ranged in size from just one-half to one acre with maximum elevations ranging from fifteen to thirty-two feet. In contrast, Nonsuch Island was a national nature reserve that was both much larger at fifteen and a half acres and much higher, with elevations up to sixty feet.[9]

A program to translocate cahow chicks from the smaller

islets to Nonsuch Island moved forward in high gear, spearheaded by Madeiros. Between 2004 and 2008, 105 chicks were
translocated from the tiny nesting islets to Nonsuch Island.
They were hand-fed squid and fish until fledging, with a remarkable 102 surviving to leave the island. The stakes were
very high to do everything right, and the success speaks to the
great care that Madeiros and his team applied to the project.[10]
To encourage these translocated chicks to nest in the artificial
burrows, Sue Schubel set up a sound system that broadcast
cahow vocalizations that were recorded by the Macauley Library of Natural Sounds at the Cornell Lab of Ornithology.

In 2008, the first of the Nonsuch translocated chicks returned as adults. The big news followed in 2009. Equivalent to
the puffin with fish at Egg Rock, Madeiros confirmed the first
breeding of cahows on Nonsuch in 2009—the first breeding
at the island in about four hundred years! In 2013–2014, Madeiros tallied at least 107 nesting pairs, and 13 of these are now
nesting at the restored colony on Nonsuch.

Spreading the nesting from three acres of storm-battered
islets to the fifteen acres of Nonsuch is important, but with the
encroachment of climate change and constant threats of development, a more regional assessment was also necessary. By
attaching geolocators that measure day length and time of the
year to cahow leg bands, Madeiros and Nicholas Carlile discovered that parent cahow were flying up to 450 miles per day
to reach food-rich waters off western Europe, especially an
area near the Azores of Portugal. Others were traveling north
to the Gulf of Saint Lawrence and the Grand Banks of Newfoundland to obtain food for their chicks in Bermuda! About
a third of the birds stayed closer to home within the area
of Bermuda, Nova Scotia, and North Carolina throughout
the year. The tracking devices demonstrated that, over twelve

months, some cahows flew more than sixty-two thousand miles, the equivalent of more than twice around the Equator![11]

Their results reminded me of the winter travels of puffins gleaned from three birds that carried geolocators. The puffins traveled north after nesting in the Gulf of Maine to the Gulf of Saint Lawrence and Labrador Sea and then south to the Bermuda plateau before returning home to Seal Island. Ironically, Maine puffins might mingle with cahows at sea!

Reflecting on how the cahow nearly became extinct, Wingate said, "We have to remember that these systems evolved for hundreds of millions of years. On Bermuda, the cahow had no defenses when the first humans set foot five hundred years ago. We're not going to save nature without human intervention. As much work as we've done on Nonsuch, all it takes to set us back is somebody releasing a dog on the island and letting it run rampant. One day, we'll probably have to try and establish the cahow on an island that is even more distant from cats, dogs, and rats."[12]

Natural events such as volcanic eruptions can cause extinctions, but this would be unlikely without humans tilting the odds against the birds. On Japan's twelve-hundred-foot-elevation Torishima (Bird) Island, 370 miles south of Tokyo in the Izu Islands, the short-tailed albatross is making a dramatic comeback. Here this magnificent bird had a very close call with extinction. Once the most numerous albatross of the North Pacific, numbering more than five million birds just a century ago, the short-tailed albatross was thought to be extinct by 1949 owing to relentless feather hunters. In 1951, twenty-five of these beautiful birds with golden heads and pastel blue and pink beaks were "rediscovered" nesting on Torishima Island.[13]

But Torishima is an active volcano, erupting as recently

as 1902, when 125 people who were living on the island engaged in the feather industry perished. The surviving birds chose to nest on the steep volcanic slopes probably because they were less accessible to the feather hunters. A landslide triggered by an eruption could wipe out the tiny colony. Luck has generally been with the surviving birds. There have been no recent eruptions, though many have perished from landslides. By the early 1990s, the number of birds had crept back to between three hundred and four hundred, but frequent mudslides have taken their toll, obliterating nests and eggs, sometimes even burying these massive birds while they were incubating.

In 1992, our success using social attraction to repopulate Maine islands with puffins and terns gave Japanese ornithologist Hiroshi Hasegawa an idea. He interested the Japanese ministry of the environment and the Yamashina Institute for Ornithology in using social attraction in an attempt to attract some of the birds to nest in a level area on Torishima Island, where they would be less vulnerable to landslides. Japan's most accomplished carver of wild birds, Haruo Uchiyama, carved the originals of two postures that were replicated into a flock of fifty rugged plastic decoys. Unlike my simple models that sat on a single post, his realistic decoys displayed detailed feathers and glass eyes. They even had two legs, complete with webbed feet and toenails.

In 1992, Hasegawa set out fifty decoys and played the recordings on a part of Torishima that was less prone to an eruption and landslides. The following year, 1993, forty-five additional decoys were produced and set in place. As in my own project with puffins, Hasegawa needed great patience. In the third year of the project, a single pair nested among the decoys. Much to everyone's excitement, the pair returned to the decoy flock for many years and fledged a chick in most years. But the albatross recovery team and all those that hoped for a rapid response were disappointed when, year after year, the single pair was alone with the decoys. It took a decade for the next pair to settle into this location. Among the albatrosses joining the new colony were several that were progeny of the first pair.

In December 2014, Hasegawa returned to Torishima for his 116th trip and was elated to count 488 breeding pairs in the main colony that occupies the ash-covered slopes and cliffs on the volcano, while the new subcolony on more stable surface started with decoys and audio recordings had grown to 183 pairs. The new colony now constitutes more than a quarter of the 681 nesting pairs. Still, Torishima is less than a mile in diameter, which leaves most of Japan's short-tailed albatrosses one eruption away from extinction.

Other nonvolcanic threats made Hasegawa worry about the bird's long-term survival. As early as 1993, Hasegawa wrote, "I was exasperated to find that 75 percent of the young birds examined had plastic in their stomachs. Perhaps, parent birds fed plastic to their babies, believing it was food."[14] Hasegawa predicted that, short of a massive volcanic eruption, the colony will increase to five thousand birds by 2018.

The only other principal short-tailed albatross nesting site is on the Senkaku Islands, but here the ownership is disputed by Japan, China, and Taiwan, and researchers rarely have access to monitor the small colony. There is new hope for the short-tailed albatross, however, from an exciting chick translocation program headed by Tomohiro Deguchi of the Yamashina Institute for Ornithology. He is heading a program that translocates chicks from Torishima Island to Mukojima Island in Japan's Bonin Islands, a distance of 220 miles southeast of Torishima. Mukojima is a nonvolcanic island and a historic home of the short-tailed albatross. They nested here through the late 1800s until the colony was slaughtered for feathers to stuff mattresses and pillows.

Techniques for translocating albatrosses were developed first with Laysan and black-footed albatross chicks. After methods were in place, ten short-tailed chicks were collected at Torishima and airlifted to Mukojima in 2008. From the beginning, the release site was enhanced with thirty realistic decoys, nine artificial eggs, and audio recordings of short-tailed albatross to further replicate a realistic setting. With great care, biologists hand-fed the downy chicks, successfully fledging sixty-nine.[15]

The program had some early success in 2010, when three of the nine black-footed fledglings came back to Mukojima and began courtship dances. Satellite-tracking devices attached to some of the birds showed that they had flown thousands of miles around the North Pacific, including the waters off California, Oregon, and Washington. By 2011, at least seven of the translocated short-tailed albatrosses had returned to Mukojima, and some performed courtship dances.

There was much celebration in November 2012 when a pair of short-tails built a nest and laid an egg at the Mukojima release site. One of the parents was a translocated bird from Torishima in 2008, but its mate was a surprise—an unbanded, mystery bird that was apparently attracted by the activity at this long-extinct colony. The pair produced an infertile egg in 2012 and 2013, but the fact that there is a breeding pair so early in the project is certainly encouraging. Eventually they or another pair will parent the first Mukojima chick. By early 2014, researchers had resighted nineteen of the sixty-nine translocated short-tails at Mukojima, and optimism about a successful breeding in the near future is very high.[16]

The very recent successes with two critically endangered seabirds, the cahow in Bermuda and the short-tailed albatross on Torishima and Mukojima Islands, give further support to the value of translocation and social attraction to help species that share the puffin's life strategy. This is exactly why I wanted to invest time and scarce conservation dollars to benefit Maine puffins when millions of puffins lived elsewhere. It was precisely their abundance that let me take the risks that have now proved so important to critically endangered seabirds such as short-tailed albatross.

Out of the blue one day in late May 1994, I received a call from Harry Carter, a California seabird researcher with a passion for murres akin to my own for puffins. At the time, social attraction had only been used on the East Coast for puffins and terns, but Harry had ideas about how the method might be adapted for common murres, eighteen-inch-tall auks that resemble diminutive penguins. He asked if I would be interested in hearing more about the plight of the murres of Devil's Slide Rock.

Soon thereafter, I was on a flight to San Francisco to see how I could help. I was met by Jean Takakawa from the Fish and Wildlife Service, who drove me to the overlook near Pacifica on US Route 1 called Devil's Slide, so named because the road here is so precarious that sections often slide off into the Pacific. Here we met Harry and together looked down at Devil's Slide Rock and began thinking about how to revive the once-thriving colony.

It was a relatively calm day, but white water was foaming around the rock, and breakers were heaving against the jagged shore. Harry explained that, on January 28, 1986, the *Apex Houston*, a barge being towed from the Shell Oil refinery in

Martinez, California, to Long Beach, spilled more than twenty-five thousand gallons of crude oil. This spill killed about 9,000 seabirds, including 6,300 common murres. After the spill, a colony of about 2,900 murres abandoned the jagged spire. Nearly a decade had passed without nesting murres.[17]

One of the most obvious challenges for this project was the enormous surf that typically pounded the rock. On a good day, the seas kicked the rock about a third of the way to the top, but in big storms, only the nesting place at the very top of the stack was dry. The rock was certainly an ideal place for seabirds because it was so difficult for people or four-legged predators to reach. From a distance, Devil's Slide Rock looked like a forbidding, snow-capped peak. Closer inspection showed that, besides the crashing seas, the climb to the top would be formidable from the crumbling, overhanging rock, covered in slimy seabird excrement and annoying biting flies.

The notion of applying social attraction to this sea stack was daunting to say the least, because the very nature of the method is gear-heavy. If we were to take on this project, it would mean somehow hauling hundreds of life-sized decoys to the top along with decoy chicks, decoy eggs, mirrors, and audio playback systems that required car batteries, photovoltaic panels, and boxes to contain the delicate gear against the extreme conditions.

To assess the likelihood of pulling this off, calls were made to helicopter operators to see what it would cost to do an airdrop of the gear. It was soon clear, however, that they wanted nothing to do with the risky plan.

After much wrangling with lawyers representing interests of the oil shipper, there was a settlement of $6.4 million. Of this, about $4.9 million was specifically dedicated to restoration of common murres to central California with a focus on bring-

ing the murres back to Devil's Slide Rock. This settlement occurred in 1994, eight years after the accident. A bold plan was
assembled under the supervision of Fish and Wildlife, and I
was part of the team to help design and implement the fieldwork. This was the first application of social attraction for
common murres and the first time to test the methods on the
West Coast. Would the murres respond to the attraction program? I called on Sue Schubel, who joined Project Puffin in
the mid-1980s, to help hatch a plan to bring the murres back.

Many people have helped the Devil's Slide Rock project
over the years, but key to early success was Harry Carter, who
was not only well informed about the biology and conservation history of California murres but a fearless Zodiac operator who could maneuver the rubber craft in and out of the
huge swells to drop the biologists on a wave-washed ledge that
led to a rappelling rope. The rope was secured by the project's
first director, Mike Parker, who was as fearless with climbing
as Harry was with a motorized Zodiac.

On January 7, 1996, ten years after the *Apex Houston*
spill, we set out to climb the rock and install the first of 380
life-size murre decoys, two solar-powered compact disc players, and several boxes with mirrors. To these "standard" social
attraction tools, we added decoy chicks and decoy eggs. We
wanted to do everything possible to tilt the odds in our favor.
Landing all of this and hauling it to the top of Devil's Slide
Rock was a herculean task, but given the size of the settlement
fund, there was a huge incentive for success.

As usual, there were skeptics. I was accustomed to this
aspect of a new project after persisting against Ralph Palmer's
ridicule of the Egg Rock puffin restoration. Of course skepticism is a key part of science, because it puts the onus of proof
onto the scientist who proposes a certain hypothesis. Repli-

cation is the best proof of a hypothesis, so I was keen to see whether social attraction would unfold as I proposed.

For the project to succeed, I had to define, up front, what I meant by success. It also meant going the extra steps from beginning to end, never settling for something if we could do better. For example, we worried for weeks about the paint that we would use on the decoys. Would it flake off in the hot California sun? Was it the right color and sheen? Were the decoys too black, when murres are actually more of a bittersweet chocolate color? Later, we would laugh at our early worries about the color of our decoys, since seabird guano soon coated everything in a ghostly chalk-white hue.

I predicted that we would be fortunate to have murres simply land with the decoys in the first year of the project. Observing courtship would be a great bonus. I assumed that it would take years before prospecting would lead to the first egg. But I was soon to receive a shock. Within forty-eight hours of putting the decoys in place, "Seabird Sue" called with the extraordinary news that a murre landed the day after the installation. The next day, less than forty-eight hours after the decoy installation, four murres were exploring the island. And even more remarkable, on May 26, the first of six pairs laid an egg. Five additional eggs followed, and three chicks successfully fledged.[18]

The surprising speed of the colonization most likely came from returning survivors from the original colony. Perhaps they had nested here before or they were chicks hatched at the colony but were still maturing at sea when the oil disaster occurred. Because they were not banded, we will never know the age or origin of these first colonists, but it is likely they were previously nervous about nesting alone where they would be exposed to predators and found the necessary confidence to nest from the artificial social scene of decoys and audio. Sue worked on the project for two years and was joined in her second year by Jennifer Boyce, who, like Sue, previously served as supervisor of Eastern Egg Rock. Jennifer's passion for seabirds is great, but a rough landing and cracked kneecap in her third year of leaping off fast-moving Zodiacs gave reason to consider using her skills and experience on a larger stage. She now holds the position of program manager for the Montrose Settlements Restoration Program of the National Oceanic and Atmospheric Administration (NOAA), where she oversees and designs projects to restore seabirds and other wildlife damaged by chemical pollution. She has found ways to stay engaged with Devil's Slide Rock through her role as NOAA's trustee council member on the Apex Houston Trustee Council. She serves a similar role on the Luckenbach Trustee Council, which now provides funding for continued monitoring of Devil's Slide Rock.

"Looking back at the use of social attraction at Devil's Slide Rock," she explained, "it was more complicated than anyone could imagine. It was a sea stack in the ocean, with no vegetation, just a flat area on top of the rock. It was covered so thick with kelp flies that they would cover your sandwich if you gave them a chance. The spray zone was beyond slippery and the swells came up so high, you'd have to run down onto the slippery rocks to jump in the boat. If you didn't make it, you'd have to run back up the rocks, wait for the swell to recede and then try again. We had amazing boat drivers who dragged me out of the water many times. That water was so cold!"

The decoy and audio program continued until 2005. By 2007, there were four hundred breeding murre pairs on Devil's Slide Rock, and by 2013 the number surpassed two thousand, about the number nesting on the island before the oil spill. The current director of the program, Gerry McChesney, concludes that we can now consider the Devil's Slide Rock colony "largely restored."

This project is notable in that it is the only project with which I have been involved that is almost self-sustaining. The murres have increased to a point where they can hold their own without management of habitat or predators. Yet other human disruptions are not far away. The Devil's Slide Rock murres cling to their tenuous home surrounded by hovering helicopters, private aircraft, fishing boats, and a burgeoning population of predatory ravens that, like gulls on the Maine coast, benefit from garbage and the excesses of a human-dominated coast. Likewise, fluctuations in forage fish in response to fisheries and climate change are ongoing issues that can keep murres from reproducing for years at a time. A new threat is the increasing numbers of brown pelicans that can displace incubating murres, exposing eggs and chicks to predators. And of course there is the ever-present threat of another oil catastrophe.

For now, the success at Devil's Slide Rock is a refreshing example of people stepping in for restoration and then backing off to follow the outcomes.

I had no sooner begun to reduce my involvement with the Devil's Slide Rock project when another West Coast seabird controversy brought me back to the Pacific. During the early 1990s biologists began noticing the growth of a huge Caspian tern colony in the Columbia River estuary. By 1997, about 7,500 pairs of Caspian Terns had moved to an artificial dredge spoil island known as Rice Island that was created by the US Army Corps of Engineers.[19]

These terns were apparently moving to Rice Island from colonies as far south as central California and as far north as the border with Canada. It was a mystery why such a big colony had formed there until diet studies revealed that they were feeding largely on young salmon and steelhead (salmonids), which were passing Rice Island during their migration to the Pacific Ocean. Despite hatchery programs to rear and release hundreds of millions of young salmonids (smolts), most species of salmonids from the Columbia River Basin are threatened or endangered because of overfishing, habitat destruction, and a system of huge dams and their slack-water impoundments that impedes adults and young alike in their migrations to and from their upstream spawning grounds.

Most of the young salmonids that migrate past Rice Island were raised in hatcheries; some of these were barged or trucked past as many as eight massive hydroelectric dams on the lower Snake and Columbia Rivers and released below Bonneville Dam, the last dam on the Columbia River, with the hope that they would continue downstream to sea, mature, and return to the river years later. But the terns had a different idea. It was as if they thought this was a feeding program for their benefit. The Rice Island Caspian terns feasted on millions of the smolts, each of which cost the federal, state, or tribal hatcheries about thirty-five dollars to raise.

Of even greater value, however, were the threatened or endangered smolts, a priceless treasure. Biologists estimated that in the late 1990s the Caspian tern colony on Rice Island was consuming about twelve million salmonid smolts per year, which represented about 10 percent of the young salmonids that were migrating out of the river. And each endangered or threatened salmonid eaten by the terns brought the fish closer to extinction. Combined with losses from the dams, overharvest, and deteriorating freshwater habitats, prospects for the endangered salmon had become increasingly grim.[20]

This was classic double jeopardy. The National Marine Fisheries Service, which is charged with protecting and restoring the endangered and threatened fish, was outraged by the losses. They called for eliminating the tern colony and set about a plan to haze the terns, destroy their eggs, and even kill adults, if necessary. The fact that the fish were now threatened or endangered primarily by overharvesting, habitat destruction, and the giant dams on the Columbia and its tributaries was usually considered irrelevant. A tern with an expensive or endangered smolt dangling from its beak was too conspicuous to ignore.

Yet the concentration was bad news for the terns as well, because the Rice Island colony had increased to nearly nine thousand breeding pairs by 1998, which represented about two-thirds of the species' population in western North America. This was now the largest Caspian tern colony in the world. A disastrous storm, disease, toxic spill, or predator attack would have obvious terrible effects for the tern population.

Ironically, the Caspian tern versus salmon controversy pitted two of the nation's most important wildlife conservation laws against each other. In defense of the terns, the National Audubon Society and other conservation groups took the Army Corps of Engineers to court over plans to disrupt the colony. The terns and its eggs were protected by the Migratory Bird Treaty Act of 1918, the primary law that protects most North American birds. The rare salmonids were protected by the Endangered Species Act.

My involvement began with a call for help from Dan Roby, whom I had first met in Yellow Springs, Ohio, where he took my field ornithology class in 1970. Dan was already an enthusiastic birder as an Antioch undergrad. He had stuck to his birding interests and obtained a PhD at the University of Pennsylvania, where he had conducted studies with least auklets.

Dan is now wildlife unit leader for the Oregon Cooperative Fish and Wildlife Research Unit and professor at Oregon State University. Usually mild-mannered, Dan was being swept up in the controversy that was raging around Caspian tern predation on salmon. As the issue was being battled out in court, Dan hatched a plan to attract the terns off Rice Island to recolonize East Sand Island, a historic Caspian tern nesting island located about fifteen miles downstream, closer to the open ocean. The idea rested on the hypothesis that, closer to the ocean, the terns would find a more varied food supply of marine forage fish, such as anchovy, smelt, and herring, and they would feed themselves and their young on these alternate foods, permitting the salmon smolts to slip past the bottleneck at Rice Island and reach the ocean.

My role was to help Dan convince the players that his "wild idea" had a chance. And as usual there was no shortage of naysayers who said the idea could not work. These included some of the best-known seabird biologists on the West Coast, who argued that even if the birds did move off of Rice Island, they would just fly back up the river to the "meat bucket" in the swirling waters around Rice Island for the easy eating. They argued that Caspian terns migrate to Mexico and beyond in the winter and that surely a short foraging trip upriver a few miles would be worth the effort to score an easy meal of plump, hatchery-raised salmon. We were told that hoping to move the birds was naive, with little chance of any real benefits to endangered and threatened fish. I countered that we knew that terns were highly attracted to social signals and that we had started more than a dozen tern colonies in Maine. Nothing tried, nothing gained. We received a hesitant green light to proceed with a pilot project.

In early March 1999 the project at East Sand Island was launched. The US Marine Corps was first to arrive. Eager to practice landing heavy equipment on a beach, they brought several large earth movers with broad blades ashore and began scraping away the accumulated weeds that had overgrown the sand where Caspian terns had nested fifteen years earlier. Then our signature tool was brought to the island: life-size Caspian tern decoys crafted from polyethylene by Jim Henry of Waitsfield, Vermont. Jim was a pioneer in the craft of rotomolding, which he used to build the well-known brand of Mad River Canoes. Now he uses his skill at rotomolding through his company, Mad River Decoys, to build rugged bird decoys for conservation (it amazes me that social attraction has grown to such proportion that a company could thrive on the demand for decoys).

Nearly four hundred decoys and several sound systems built by Seabird Sue were installed at East Sand Island by April 1. To discourage the terns from nesting at the Rice Island colony, the Army Corps of Engineers planted a crop of winter wheat over most of the area formerly used by nesting terns. Where the wheat did poorly in the sandy dredge spoil, rows of three-foot-high silt fencing were erected as a deterrent for the nesting terns. The idea was to degrade the nesting habitat to make it look unsuitable for the beach-nesting terns.

At the time, Dan and his colleagues were trapping some of the terns by using cannon nets, this device is shot over roosting birds. The netted terns were quickly taken from the nets and banded; some also received satellite transmitters to follow them over the course of the season. This work occurred early in the nesting season, soon after the terns arrived at the colony.

As we hoped, some terns found the decoys on East Sand Island, heard the recorded courtship calls, and settled in to nest on the expansive sandy habitat. The rest soon followed. By the beginning of the third year (2001), nine thousand pairs of Caspian terns had moved to East Sand Island from Rice Island. Much to Dan's relief, the feeding studies that his team performed confirmed his prediction that the terns at East Sand Island would feed on alternate foods obtained from nearby water, shifting from a diet of 74–90 percent salmonids to just 33 percent. At East Sand Island, the balance of the chick diet was other readily available marine forage fish such as herring and anchovies.[21]

Dan is now one of the main architects of a multistate initiative to move most of the nesting terns out of the Columbia River estuary to benefit salmon smolts and spread the terns to many nesting sites for their own safety. Too many eggs in too few baskets applies to seabirds everywhere. With this in mind, social attraction is the primary tool now used to redistribute Caspian terns to alternate nesting locations throughout the Pacific Northwest. This includes encouraging the birds to colonize nesting islands in inland lakes and reservoirs. Some of these efforts involve impressive uses of barges to create artificial islands, a promising approach that helps give the birds new habitat while reducing risk from terrestrial predators such as raccoons, foxes, and coyotes.[22]

In the court confrontation between conservation groups and the corps, the birds won on a technicality. A federal judge decided that the corps had not complied with the National Environmental Policy Act when it decided that an environmental impact statement (EIS) would not be required to remove the

Caspian tern colony on Rice Island. The court concluded that a thorough EIS needed to be conducted, and out of this grew "The Caspian Tern Management to Reduce Predation of Juvenile Salmonids in the Columbia River Estuary Final Environmental Impact Statement, known as 'The Caspian Tern Plan.'"

The success of restoring and creating new Caspian tern colonies in the Pacific Northwest positioned Dan to assist one of the most important seabird recoveries on earth. The Chinese crested tern, arguably the world's rarest seabird, is teetering on the edge of extinction, with fewer than fifty individuals remaining in the East China Sea.[23] The main threat to the species is egg poaching by fishermen, so officials in the People's Republic of China are attempting to establish a colony of the rare terns on an island sanctuary that will be safe from egg poachers along the Chinese coast. Social attraction is the centerpiece of this ambitious recovery plan in the Zhoushan Archipelago off the coast of Zhejiang Province.

During the summer of 2013, three hundred decoys and two sound systems constructed by Seabird Sue were installed on a small island called Tiedun Dao with the hope of ultimately luring some of the remaining Chinese crested terns to a location where they would be safer from egg collectors. It was hoped that the still-abundant greater crested terns would initially colonize the island, their numbers would then gradually grow, and Chinese crested terns, which have always been found nesting within large colonies of greater crested terns, might eventually follow. This was a very familiar relationship to me, since the endangered roseate terns of the Maine coast always nest embedded within large colonies of common terns.

The Chinese crested tern restoration team expected that it would take several years before there was any hope of attracting the birds to the island, but they were amazed by the results. In the first year of the restoration effort, hundreds of pairs of greater crested terns nested on Tiedun Dao and raised several hundred young. At least two pairs of Chinese crested terns nested with the greater crested terns, and one pair succeeded in fledging their single young tern.[24]

The project continued in 2014 with a resident team in place on the island. Much to their amazement, about twenty-one pairs of Chinese crested terns flocked to Tiedun Dao along with about two thousand pairs of greater crested terns. It appears that nearly all remaining Chinese crested terns are now nesting on this small island. This is good news because the few remaining birds have a better chance of successful breeding, but all of the eggs are still in this one "basket." The attraction and protection model is working; an egg poacher who approached at night was spotted and arrested after the resident tern stewards alerted authorities.

In 2003, a New Zealand Department of Conservation global report on transplanting chicks to restore a species—aka "chick translocation"—reported that our efforts on Eastern Egg "established a baseline for seabird translocation techniques." The 2010 State of the Birds report, a collaboration of the US government and top conservation groups including Audubon, said translocation was among the "key steps" of conservation for birds that will continue to be threatened by climate change, commercial fishing, development, and pollution.[25]

Despite these acknowledgments, social attraction is much more commonly used than the much more labor-intensive and expensive process of chick translocation. The principal advantage lies in the chance for almost instant success, since the method focuses on attracting breeding-age adults, rather than working with chicks that have a slim chance of survival and then may take half a dozen years to reach breeding age. Social attraction works best with species such as terns that naturally move from colony to colony, looking for a suitable nesting site, perhaps even a year or more before they actually commit to laying eggs; it is less likely to succeed with species where the young tend to return to the place where they hatched.

I learned this information the hard way with gannets. These great white northern boobies are a classic case of a species that rarely pioneers new colonies. Only six colonies presently exist in all of North America, mostly in the Gulf of Saint Lawrence. In 1995, confident from many successes with terns, Galápagos petrels, and murres, I was eager to test some of the limits of the social attraction method and set out on an ambitious project to establish a gannet colony in Maine.

I started by looking at recent history. Northern gannets bred in the Gulf of Maine at several sites until the feather-hunting days of the late 1800s. There were even islands named for gannets: Gannet Rocks near Machias Seal Island in New Brunswick and Gannet Rock near Yarmouth, Nova Scotia. I had powerful memories of visiting gannets at Bonaventure Island, located off the tip of the Gaspé Peninsula of Quebec, during 1967 and 1968, the same summers that I first visited Machias Seal for puffins. Once again, I was imagining a restored seabird colony. But this time it was gannets nesting on the sea cliffs of Seal Island National Wildlife Refuge. We were already well established on the north end of the island with a field camp protecting puffins and terns, so it seemed logical to test social attraction for gannets within sight of the field camp on the steep sea cliffs of the island.

I started by ordering twenty-three life-sized polyethylene gannet decoys from Jim Henry at Mad River Decoy. Of course this was not something that he already had on the shelf, so Jim had to create an original in two postures (incubating and alert), just as Donal O'Brien had created models for the first tern decoys. In early May we set out the decoys in a likely place and did everything we could think of to make them look and sound like the real thing. To replicate an authentic colony, we sat the decoys on mounds of seaweed and even painted the rocks with whitewash to look like guano. By this point I was convinced of the importance of audio recordings, so we set up a solar-powered sound system and played recordings from the Bonaventure Island gannet colony. Everything looked good to us. But in 1995 and 1996 we watched without seeing a single gannet land among the decoys.

I eventually concluded that although there were plenty of gannets near Seal Island, these birds probably were not interested in nesting here because they had strong memories of their natal home in Canada. Unlike the puffins that we had translocated from Newfoundland as chicks and that were given abundant opportunity to learn that Maine was home, I had no real reason to believe that that gannets (notable for their faithfulness to their natal home) would be easily duped into nesting in Maine. This tendency (called philopatry, from the Greek *philo*, loving, plus *patris*, of the father) explains why there are presently only six gannet colonies in North America. The nearest of these to Seal Island was Bonaventure Island.

The gannets' tendency toward philopatry and complete lack of success at Seal Island led me to conclude that we needed to be closer to a source colony, so we packed up the decoys and sound system from Seal Island two years after starting the project. In collaboration with Kathy Blanchard, now with the Quebec Labrador Foundation, and Richard Sears of the Mingan Island Cetacean Study, I hatched a plan to move the gannet decoys and the audio system to Perroquet Island in the Mingan Islands of Quebec—an island with a history as a gannet-nesting site.

Perroquet Island (named for its colony of sea parrots, that is, Atlantic puffins) is located on the north shore of the Gulf of Saint Lawrence. An "immense" colony of gannets nested on the island until 1859, when excessive hunting of the birds for cod-fishing bait followed by disturbance associated with the construction of a new lighthouse eventually ended the era of gannet nesting.

In 1997 we found the lighthouse, now part of the Canadian national parks system, automated and abandoned. I had thought that our idea would be warmly embraced, but on the contrary, some of the park staff worried that the gannets would overrun the island as soon as we put out the decoys and would chase off the puffins, which brought income for local ecotours. There was also worry by some that a gannet colony would interfere with plans for a museum in the lighthouse.

These reservations aside, we obtained permission for a pilot study in 1997 and placed forty-seven gannet decoys and an audio system on the island. The decoys were now much closer to gannet colonies in the Gulf of Saint Lawrence. Although about two hundred pairs nested on a tiny island off the western tip of Anticosti Island, the famous colony on Bonaventure Island (largest in the world) was my greatest hope for a source of colonizing gannets. With Anticosti and Bonaventure Islands located fifty and one hundred miles, respectively, to the south of Perroquet Island, this strategy seemed a great advantage over Seal Island off the Maine coast, which was about three hundred miles south of Bonaventure Island.

I was hopeful that increased numbers at the Bonaventure colony would increase prospects for success since this population had grown in recent years to about fifty thousand pairs, in part due to a notable decline in residues of the pesticides DDT and dieldrin in eggs.[26] Soon after we started the project, it was clear that Perroquet Island was more attractive than Seal Island.

After two years without landings or even close fly-bys, we were thrilled to see as many as three gannets landing at one time among the decoys. By 1999, in the third year of the program, a pair of gannets spent most of the summer together among the decoys and was observed pulling vegetation and tucking it under themselves in nest-building behavior. While this all gave us hope, these were also the most encouraging sightings in five years of gannet attraction. By 2001, gannet sightings at Perroquet Island were the lowest recorded to date, and a decision was made to suspend the program while Parks Canada completed their planning for the island.

I began thinking about ways to ratchet up the effort and concluded that translocating gannet chicks would be necessary. But this would be much harder than translocating puffins because the gannets nest on the surface of the ground and I worried that chicks might become imprinted on their human foster parents. Also, the thought of feeding dozens of ravenous young gannets with a surrogate gannet puppet (to reduce risk of visual imprinting) was a daunting logistical worry. Perhaps if we had stayed with the program longer, we would have met with success, but this was a stand-alone project rather than an add-on project (for example, adding terns to a puffin colony project). Here we lacked both financial support and encouragement from Parks Canada to continue.

The most important lesson learned with the Seal Island and Perroquet Island gannet projects is that social attraction works better with some species than others. It works for terns in part because they have a weak tendency to return to their natal home. Most terns naturally move from one island to the next in response to predators, overgrowth of weeds, and habitat loss from coastal erosion. In contrast, gannets nesting on high cliffs are safer from such problems and return to the same nesting sites year after year, racing back to get the prime sites at home rather than going off and pioneering new colonies that would tend to be small and at high risk for predators. Nesting in large colonies also helps gannets share locations of schooling fish. The gannet's keen eyes can certainly spot at a great distance the glistening white backs of other gannets from the colony plunging into the sea in pursuit of mackerel, herring, and other schooling prey. For these reasons it makes sense for gannets to nest in large colonies and to avoid the risks of pioneering new colonies. Little wonder there are just six colonies in North America.

Just as I was beginning to believe that I could think like a gannet, I received news in January 2010 that two Australasian gannets had laid eggs among decoys and audio equipment at Young Nick's Head, New Zealand. This news was of great interest because the Australasian gannet is very similar to the northern gannet of Canada in appearance and in its colonial nesting behavior.

The new colony was located twelve miles from Gisborne City at the tip of a wild peninsula about a hundred feet above the sea. Here, New Zealand biologist Steve Sawyer placed eighteen fiberglass gannet decoys on seaweed nests with whitewash, similar to the procedure we adopted on Perroquet Island. Before his success at Young Nick's Head, he had tried for two years at the tip of the Mapiri Peninsula several miles to the south. Here the decoys were clustered and placed into a protected swale at the end of the peninsula. Nearly two years passed without a single gannet landing.

Thinking that the swale may have prevented passing gannets from seeing the decoys, Sawyer moved the project to the tip of Young Nick's Head, a jutting promontory just four miles north of the Mapiri Peninsula. That shift made all the difference.

When I asked Sawyer for more details, he explained that even from the outset in 2008, the Young Nick's Head colony was off to a promising start when a gannet swooped close for a look just minutes after the audio system was clicked on. By January 2009, two hundred birds were roosting at the site, and a year later the first eggs and chicks made history. By January 2014, at least eighty-three gannet chicks were recorded in the Young Nick's Head colony. Apparently, once those first brave pioneering gannets had made the commitment to the artificial colony, they became living decoys that helped to attract other gannets. Now the new colony was behaving much as a growing tern colony. It was beginning to look as if gannet colony creation was mostly about location—near a thriving source colony and in a conspicuous place to display decoys to lure passing birds.[27]

To make the gannet colonization process even more confusing, a single pair of northern gannets nested on Whitehorse Island, New Brunswick, in 1999, and another (or possibly the same pair) nested on Machias Seal Island in 2012.[28] Both nesting attempts occurred without decoys or sound recordings. Apparently, these were examples of birds without the usual inclination to return and nest at their natal colonies—evidence that occasionally birds are just wired differently in this aspect of their behavior.

Were these gannets dysfunctional, confused, and maladapted for survival, or does this behavior outside of the norm prove vital to the species over time? All of these descriptors likely apply. How else could one explain how such a philopatric species could ever found a new colony?

Not surprisingly, both attempts failed to fledge chicks, as these pioneers faced the severe predation risks of nesting away from existing colonies. The pair on Whitehorse Island raised a chick for several weeks, but it was likely taken by an eagle; the egg on Machias Seal was probably a snack for a hungry gull.

These nestings and the outcomes also point to the importance of chance events. That two birds with a weak sense of philopatry would find each other hundreds of miles from their own natal colony is notable enough. But to actually produce an egg, hatch it, and raise a young bird alone from other gannets for even several weeks is remarkable. If they had succeeded in rearing the chick and returned to do the same in subsequent years, it's likely they would have been joined by others—just as our decoys tend to attract feathered birds. Such small colonies are especially vulnerable to chance predation or weather events, but with luck one can see how a colony might persist and with time grow to a substantial size, assuming that there was ample food in nearby waters.

There are many reasons why social attraction is used more often than translocation of seabird chicks. Although it can take many years or end in failure (as the northern gannet example shows), it is still a shortcut when compared to a chick translocation project. It also appeals to those concerned with the genetic effects of bringing chicks from remote areas into habitats where just a few surviving birds remain of a local genetic strain.

But those who worry about the dilution of local genetics with such chick transfers should consider that there is a greater concern of losing a local population or even a species by taking the path of wait and see. To do nothing when a species (or an isolated population) is down to the last few individuals usually results in extirpation because chance events of predation or other disaster are greater on a small population. In contrast, larger, more robust populations can withstand attacks from predators, storms, disease, and other avian disasters.

New applications for social attraction will certainly be found in the future, especially with land birds. Suggesting the potential for even broader application, social attraction is now used for attracting purple martins, bobolinks, black-capped vireos, and tricolored blackbirds.[29]

The increasingly broad application of social attraction for conservation suggests that decoys and audio playbacks can help to attract even highly territorial species. Apparently, territorial birds such as vireos are attracted to social cues not for battle but to nest near what they perceive to be occupied territories. This is likely because the presence of the same species quickly informs birds that habitat is suitable for reproduction and that appropriate mates will likely be found nearby.

13

Reconsidering the Balance
of Nature

In the course of my work with Derrick in writing this
book, his wife, Michelle Holmes, a physician and epide-
miologist at Harvard University, played devil's advocate
with a simple but piercing question: "Don't get me
wrong. Putting the puffins back is a phenomenal accomplishment. But if interns have to stay every summer on Eastern Egg
Rock, how is it a sustainable project?"

This question gave me pause regarding the word *sustainability*. In our society it seems that "sustainability" has become
so politically correct that it is reflexively invoked by any entity
seeking an environmental "stamp of approval" for its activities.
Yet the smallest community-supported farms hold no more
claim to the word than the builders of the most soaring LEED
certified (Leadership in Energy and Environmental Design)
skyscrapers. "Sustainability" rolls as easily off the tongues of
executives at fossil-fuel companies as it does off the lips of pro-
ponents of solar and wind energy.

Big-box stores appropriate the term by putting solar

panels on top of their stores even as their products are made in China, requiring cheap energy to transport the goods to market. Even Coca-Cola and Pepsi, whose products contribute mightily to global obesity, have wrestled ownership of the word from the Sierra Club, Audubon Society, and Greenpeace. In a short history of popular and often contradictory use, *sustainability* has become a muddle that often results in public apathy and government inaction over environmental degradation.

Project Puffin has taught me that, because of our human impacts, the sustainability of many species requires ongoing human stewardship. I once thought that *sustainable* meant getting puffins in Maine to a population where they could survive on their own as they did before European settlement. Four decades of work has convinced me that this is not likely to happen. That in turn raises an obvious question that I must answer for skeptics. Is there any lasting importance in a project where, if I ever stopped it, the gulls and eagles would return and wipe everything out?

I argue YES because these puffins are a living definition of modern stewardship. Even a remote jumble of boulders like Eastern Egg Rock in relatively sparsely populated Maine cannot escape human effects.

Egg Rock is an island connected to the mainland, not by a bridge, but by a web of creatures from plankton to predator. The most obvious such connections are garbage landfills and the fishing waste of passing lobster boats. These support huge numbers of gulls, and each gull is always looking for the next meal.

Another connection to burgeoning coastal humanity is that the very successes of coastal wildlife conservation and changes in land use over the past century have nurtured other voracious threats to the puffin. These threats now include native predators such as resurgent eagles, owls, mink, and river otters.

New England, once a region clear cut by farming, logging, and urbanization, has become reforested to its highest levels since the Civil War. These forests are now home to eagles, great horned owls, mink, and otter that are capable of finding their way to even the most remote island nesting seabird sanctuaries.

These predators island-hop their way from the mainland to within a few miles of restored seabird islands such as Egg Rock and then swim or wing their way the last few miles. Predators make these bold incursions because our success at rebuilding seabird numbers at Egg Rock also makes the island an irresistible source of protein. I have infinite admiration for the ability of predators to find prey and know that there is no place on the Maine coast so inaccessible that it is safe from the keen eyes and olfactory talent of predators. The more success we have in building up the seabird numbers at Egg Rock and other restored seabird nesting islands, the more predators will come—because the seabirds are food, and if there is a meal to be had, something will eat it.

I am convinced that if predators such as eagles, mink, and otter were permitted to snack at will on the small colony of puffins and terns at Egg Rock, all of our tedious efforts to revive the colony would be lost. Similarly, if we were to let down the guard for even a summer, the aggressive black-backed and herring gulls would gobble up most tern chicks. The impact of gulls and their ongoing threat to the restored puffin colonies was demonstrated in 2003 when McGill University PhD student Christina Donehower set out to study the impact of gulls on tern chicks. For most of three summers, she sat with binoculars in hand on the roof of the Egg Rock Hilton watching for evidence of predation. Before this, the extent of the predation was unknown.

In the first year of her study, we stopped all control of herring and great black-backed gulls to permit Christina to document the effect of gull predators. We were shocked to discover that 33 percent of the chicks were being snatched by the gulls. Not just any gull—but a few specialists. In the following two summers, we attempted to chase off the predators and in a few cases shot some of the specialists. This reduced the predation by about half in the following two years.[1] The haunting questions I have long pondered are: How could this system have survived without human assistance for so long before our arrival? What happened that makes it necessary to manage the birds today?

I have come to believe that increased human presence and associated enterprise not only gives advantages to some species over others but usually leads to simple biological communities with fewer species dominated by generalists and scavengers. We see it on land with gulls, crows, vultures, raccoons, pigeons, and rats replacing habitat specialists such as wood thrush and meadowlarks. Likewise, the ocean floor around Egg Rock is also a simpler habitat than in centuries past. Commercial fishing emptied nearby waters of big fish such as cod and halibut. Even invertebrates like sea urchins and starfish are now largely replaced by crabs and lobsters, because scavengers readily fill in the ecological opportunities vacated by more specialized animals. Scavengers thrive around people. Even during my time in the bay, now tempered by the perspective of four decades, I have seen the demise of giant cod and recall with remorse that I contributed to the loss by hauling in a forty-pounder off Monhegan Island while on a recreational fishing boat out of Boothbay with my parents.

So, here we were attempting to restore a bird community that had been missing for a hundred years into a bottom-up modified marine community with no knowledge of where the puffins spend most of the year. Sometimes I reflect on my simple initial idea of bringing the puffins back and wonder whether, if I had known then about the resurgence of predators, complex relationships with prey, and the need for ongoing management, I would have launched the project in the first place.

When I first started transplanting puffins, I was much more focused on the day-to-day daunting challenges that lay ahead for our tiny charges. With time, I came to marvel at the resilience and vigor of the chicks. They seemed such helpless bundles of fluff with tiny beak and webbed feet! But even at this stage, every feature had a purpose. The thick down was the ideal insulation to help them stay warm when their parents were at sea; their sturdy feet helped them roam about in their underground burrow, where they had a tidy toilet area. Yet after just a month of consuming their vitamin-supplemented diet of fish, they had transformed into rugged little seabirds, their fluffy down traded for sleek, waterproof contour feathers. No flashy colors were necessary for the fledglings, dressed in shades of gray for the year ahead.

Considering that the little puffin chicks were so dependent when we brought them to Egg Rock, it was always a huge source of wonder to see them transform completely into self-sustaining seabirds, fully equipped to plunge into the frigid North Atlantic and become as oceangoing as a fish, fully capable of fending for themselves in the oft-wild North Atlantic. Fledging time was especially remarkable. Fully feathered in an all-business cloak of black and white, the pufflings somehow know where to go and how to find food and dodge predators—all without the aid of their parents. Fortunately, they are hardwired for this abrupt transition from subterranean to marine animal. In this respect, our artificial rearing replicated the usual feeding routine of parent puffins and our puffin chicks were heading off to sea just as they would in nature.

Now, it is increasingly clear that the "sustainability" in Project Puffin lies not in finding an exit strategy but in its ability to restore seabirds to historical locales by expanding ranges and increasing the number of nesting colonies. In this way, we are reducing risk from a variety of extreme threats. Building populations to greater numbers during years of plenty also helps the population thrive during lean years or after some catastrophic event such as an extreme winter storm or havoc caused by a predator.

But sustainability also means finding ways to sustain field stations such as the camp on Egg Rock. Keeping Project Puffin interns on the seabird sanctuaries is critical. They help the rare species retain populations in the face of predation from the generalist species such as gulls that thrive in the wake of large coastal populations and the eagles that now abound in part because they consume gulls.

The importance of human presence to guard against predators was underscored by an experiment that I conducted during the summers of 2009 and 2010 using a robot at Eastern Egg Rock. I came on the idea of using a lifelike "scaregull" and put the idea out to my friend David Buchner, a technology teacher at the Dewitt Middle School in Ithaca. "Great idea," said David. "We could build you a robot," he said, even though it was a stretch for his tech-savvy kids. Aided by a Cornell robotics expert, they created a life-size robot that they named Jack because he was to live in a box, rising up at random times to scare off gulls. I called him Robo-Ranger and hoped that he would prove an innovation for scaring off gulls—perhaps the first of a small army of mechanical puffin and tern protectors.

To make Robo more effective, we dressed him up with the same clothing outfit of orange hat and yellow slicker as the interns that occasionally took a shot at the gulls. He did stand

Juliet Lamb with the Robo-Ranger at Eastern Egg Rock. In an attempt to make intern and robot look similar, Juliet donned a yellow slicker, an orange hat, and a pseudo-beard and moustache. Photo by Stephen W. Kress

up, but the mechanics soon failed in the salt air and the gulls lost all fear. This reinforced my growing respect for the gulls' ability to recognize facial features and real threats. They easily won the intelligence test on this round.

Future seabird conservationists will have not only to defend existing colonies but to find ways to move birds to safer habitats in the coming years. This will become increasingly urgent because there are already about three times more people living within seventy miles of a coastline than the global average density.[2] Here, seabirds and conservationists must contend not only with the old challenges of predators, poachers, toxic spills, development, and the occasional volcano but also with the need to anticipate broad effects from climate change.

Just as canaries were once used to test the air quality in coal mines, so birds today are warning us of degraded and dangerous environmental conditions. Disappearing eagles, ospreys, and peregrine falcons once helped to make the effects of DDT visible to us. In the same way, crashing populations of warblers now tell us that there is too much logging in rain forests in Central and South America and too much cut-up, fragmented habitat in the United States. Similarly, vanishing birds such as bobolinks and eastern meadowlarks are warning us that we are sacrificing too much grassland for commercial agriculture, which is now much more about corn-based ethanol, animal feed, and high-fructose corn syrup than it is for sweet summer corn on the cob.

Likewise, seabirds are sentinels of the ocean. They are messengers about climate change and health of fish populations because the numbers, size, and kinds of prey delivered to their chicks inform us about changes in the abundance, diversity, and distribution of coastal fish. Overfishing and climate change are altering the world's fish populations at a frightening speed unimagined when we started our project in 1973.

The African penguin (formerly black-footed penguin) is a shocking example of how rapidly some seabird populations can crash. There were likely one and a half million to three million African penguins in the early twentieth century, but by 2009, the population had collapsed to just twenty-six thousand breeding pairs. Factors contributing to the decline were competition with commercial fisheries for pelagic fish, changes in prey distribution and availability, habitat degradation, oiling, and predation.[3]

Derrick covered the 2010 International Penguin Conference in Boston for the *Boston Globe* and wrote about the findings because he had visited Cape Town the year before, walking among the birds on its famed Boulders Beach. The birds were so curious and unafraid that one of them poked at his camera bag. "I had no idea that I was staring extinction in the face," Derrick wrote. "I have the chill of realizing that unless something changes, these penguins could well join the dodo, moa, great auk or any of the large flightless birds we pushed into nonexistence."[4]

Of the world's eighteen penguin species, eleven are either endangered or vulnerable, and most face serious population declines. For some penguins, their key diet of krill is disappearing with melting Antarctic ice. Wayne Trivelpiece, an Antarctic biologist for NOAA, explained to Derrick that we should think of climate change as being a bit like your freezer. When your freezer is at thirty degrees, you have ice cubes. If your freezer is at thirty-three, you have water in your ice trays. Such minor changes are having dramatic effects on penguins.

Like seabirds in many parts of the world, African penguins suffer from a combination of human overfishing for forage fish of anchovies and sardines and warming oceans that drive remaining fish farther and deeper offshore and out of penguins' feeding range. The issue of forage fish came into the international spotlight in 2012. A team of researchers assembled by the Institute for Ocean Conservation Science at Stony Brook University, with support from the Pew Charitable Trusts, discovered that the percentage of forage fish in the world's annual catch has more than quadrupled since the 1960s and is now more than a third of the global catch.[5]

Forage fish are the ethanol of the sea. Where the vast majority of corn grown in the United States today is for fuel for livestock feed, 90 percent of the forage fish catch ends up at fish farms and as pig and poultry feed. In another study funded by Pew, researchers at the University of British Columbia calculated that the world's pigs and poultry now eat six times more fish than Americans.[6]

Although the forage fish catch is worth about $5.6 billion to the global feed market, the forage fish left in the sea are worth $11.3 billion because they are gobbled up by far more commercially valuable fish such as tuna, salmon, cod, striped bass, halibut, bluefish, haddock, and flounder. If we want the big fish to stay on our plates, the tiny fish must remain on their menu. Fisheries scientist Ed Houde of the University of Maryland Center for Environmental Science, commented for one of Derrick's columns, "I'm not sure you can ever make them sexy, but if you take out the herring, you take out the whales. Charismatic marine mammals and seabirds get all the attention. But historically, we really haven't measured the abundance of their food."[7]

Many of our most endearing wildlife species depend on menhaden and herring for between half and all of their diet. Bald eagles, ospreys, and seals depend on forage fish. Pelicans, gannets, gulls, herons, and cormorants (basically all the seabirds) depend on them. Growing recognition of the importance of forage fish is slow to come, but the paper "Global Seabird Response to Forage Fish Depletion—One-Third for the Birds" has helped to draw attention to the issue of forage fish conservation. Fifteen seabird authorities examined fourteen seabird species within the Atlantic, Pacific, and southern oceans and concluded that one third of the biomass of forage fish must be left in the seas (rather than harvested by humans) to "sustain seabird productivity over the long term."[8]

On Machias Seal Island, where I saw my first puffins in the 1960s, there are clear signs of ecosystem change. There, our Canadian partners, Tony Diamond (professor of biology at University of New Brunswick) and Andre Breton (a former Project Puffin intern), have looked at puffin chick and adult survival. Their study tells us that puffin body weight has been on a long, slow decline from insufficient food. Likewise, they documented a trend where years with reduced proportions of herring in puffin chick diet leads not only to a poor year for the chicks but to reduced survival for the adults in subsequent years. This is the first puffin study to demonstrate that adults that are in poor condition at the end of the nesting season have a lower chance of surviving the rigors of the following winter in years when herring is scarce.[9]

In Europe, puffins suffer in years when warmer waters displace forage fish near nesting islands, and in some winters, large numbers have died from extreme storms and associated starvation. In the winter of 2012–2013, 2,500 dead puffins washed up on Scottish shores, most in starving condition, and likely many more died far at sea where puffins typically spend the winter.

Most alarming, the puffin colonies of southern Iceland, home to the world's largest puffin colonies, have not had a good crop of nestlings since 2004, because sand lance, the preferred food for puffin chicks, have moved too far from the nesting habitat in response to warmer water. This has led to a dearth of suitable forage fish and tragic amounts of chick starvation. As was mentioned earlier, on the Icelandic island of Heimaey, where up to eighty thousand puffins were previously taken in an annual puffin hunt, the traditional puffin hunt ended in 2010 because of declining puffin numbers. A five-day hunt in 2013 netted just three hundred puffins. Most of these were older birds—a fact that reinforces that there are few puffins in queue to replace the oldest generations.

Although I have been hoping that Maine waters would

somehow avoid the dramatic changes experienced by European puffin colonies, the summer of 2012 saw the warmest waters ever recorded in the Gulf of Maine and has changed my thinking that all's well along the Maine coast.[10] Although puffins seem to have dodged climate effects until recently, other species have suggested that change was under way for at least the past decade. For example, the overall warming of coastal Maine waters has resulted in the range of lobsters moving northward in New England in the last decade, creating a commercial glut in Maine but reducing numbers of the crustacean south of Maine. Likewise, at least fourteen fish stocks have shifted northward in New England, and some of these have also moved into deeper water where they are out of reach of puffins.[11]

One of those shifting stocks is white hake.[12] This change is especially alarming because white hake usually makes up about 75 percent of the food fed to Egg Rock puffins. In 2012, there was a surge of more southerly butterfish in the puffin chicks' diet, which turned into trouble because the fish were too large for the chicks to swallow.

Through a video camera tucked into one of our burrows on Seal Island, we watched one chick struggle with the oversized fish and later found that it starved surrounded by butterfish. It was not alone. An end-of-the-year check of burrows found that about a third of the puffin chicks that year suffered a similar demise, and similar dismal nesting occurred at nearby Matinicus Rock. Trouble for puffins continued during the winter of 2012–2013, when more than forty washed up starving or dead on the shores of Massachusetts and Bermuda. Several were carrying Project Puffin bands. And because puffins typically winter far offshore, it's likely that many more perished far at sea.

Butterfish are round-bodied, making them difficult for young puffin chicks to swallow. This puffin is about to deliver three large butterfish and one small white hake (a more manageable food) to its chick. Photo by Derrick Z. Jackson

The 2013 season brought more questions. Fortunately, butterfish were scarce in the puffin diet, but the white hake and herring were not present in sufficient numbers to provide ample food for the puffin chicks. Although little is known about how far or exactly where puffins typically find food for themselves and their chicks, they apparently could not find enough food near the islands and either abandoned their egg early in the season or stopped feeding chicks. This led to just 10 percent of pairs fledging chicks, and those pufflings that did fledge were of lighter weight than usual—which further reduced their chances for survival at sea.

Oddly, the puffins on Eastern Egg, the southernmost puffin island in the United States, had a successful season. Here the population increased by 7 pairs to 111 pairs, while the number of nesting pairs at Seal Island and Matinicus Rock was down by a third.

Some would take the events of summer 2012 and 2013 as certain news that Maine puffins will decline in the coming years as climate change advances. Pessimists would say that this is surely similar to what is happening to puffin populations in the eastern Atlantic, especially in the Shetland Islands of Scotland, Norway, and southern Iceland. At these once-prolific puffin colonies, warming water is linked to long strings of unproductive years for puffins. At these colonies, puffin populations are declining because parent puffins cannot replace themselves. But nature is complex, and the lessons are in the details.

It is unclear whether the news coming from puffin colonies in Europe is a precursor of what will happen in Maine or whether the troubles there are uniquely related to local changes in specific fish populations. Norwegian puffin populations, for example, have been stressed for decades now because of a chronic shortfall in herring. At the Norwegian island of Røst, nearly all puffins have failed to produce nestlings in recent years. Puffins at this large colony depend on herring that drifts northward with currents into Norwegian waters. The value of this food depends on how fast the herring is growing. In good years, herring mature from larvae to decent-sized fish several inches long by the time they arrive near the nesting colonies. However, in recent poor years the herring are still in larval stage when they arrive in Norwegian waters and offer few calories to the puffins. In years when this herring conveyer belt shifts farther from the puffin nesting islands, the parent puffins cannot find enough food near their colonies—resulting in massive chick starvation.[13]

Not all the news from Scandinavian seabirds is grim. For instance, in northern Iceland, where capelin stocks are still ample, the puffins continue to thrive. But the overall concern was amplified in 2014 with reports of breeding failures not just of puffins but also of black-legged kittiwakes and Arctic terns. Researchers say climate change is likely upsetting food chains and breeding cycles.[14]

As I write these pages, it is too early to know whether the changes observed during the summers of 2012 and 2013 are the new normal or anomalies. But the 2014 puffin nesting season sheds light on the future as it demonstrated how even small changes in sea surface temperature and the related timing and intensity of the spring plankton bloom can make or break the puffin nesting season. Following the cold winter of 2013–2014, the average Gulf of Maine surface temperature dropped by almost 2 degrees Fahrenheit from the previous year, bringing it into the high end of the "normal" range. This water temperature combined with the flow from rivers filled with freshwater from heavy, abundant winter snow to create a more typical spring plankton bloom, resulting in ample herring and white hake for puffin chicks throughout the nesting season. A surprise was a bumper crop of sand lance that further boosted the nesting success.

These conditions resulted in puffin production returning to the normal range at Seal Island, with 75 percent of the pairs successfully fledging chicks. For the same reasons, the Egg Rock puffin colony surged to a new record high of 148 pairs, up 33 percent from 2013. About 750 healthy puffin chicks fledged— a bumper cohort of young puffins that we hope will begin returning in 2016. Future successful nesting cycles will take a similar alignment of air temperature, snow, and currents to set the stage for the next good year.

Just as these summers have given me pause about the sudden changes in fish populations related to climate, there appear to be some hopeful fish events on the way that may help the puffins adapt to the changes. Recovering stocks of bluefish, redfish, and haddock may fill the gap if white hake move farther north or into deeper water. I find it hopeful that Maine puffins are now flying back to their nesting islands with their colorful beaks loaded with redfish and haddock, because these are among the stocks that the Natural Resources Defense Council said are rebuilt or well on their way to being rebuilt in New England waters as a result of the Magnuson-Stevens Fishery Conservation and Management Act, the guiding legislation that shows promise of helping to rebuild depleted fish stocks.[15]

I am cautiously hopeful that Maine puffins can supplement their usual diet of hake and herring by adding other small fish if they are available to them. In 2010, haddock made up 30 percent of the food items delivered to puffin chicks at Matinicus Rock, and redfish constituted nearly 25 percent of food items in 2011. It would seem that red hake, a more southerly fish, could also become more important in the future, but

Michael Fahay, a hake specialist, points out that it spends a shorter time near the surface of the food column before it descends to the bottom, where it spends the rest of its life in deep water and is unavailable to surface-feeding terns and perhaps even deep-diving puffins. In contrast, young white hake stay at the surface and in higher parts of the water column longer until they are several inches long—a better-size meal for a hungry puffling. Such behavioral differences in fish will prove key to determining whether a future forage fish is suitable.

Atlantic herring, already a key puffin food, could become even more important if fisheries are managed in a way to leave more fish for seabirds. It is the ideal food for puffins and terns because of its high fat content and its size, shape, and habit of spending time near the sea surface when young. But herring is also the preferred bait for lobsters on the Maine coast, with about two hundred million pounds used annually. Finding an alternative bait such as animal hide or a healthy fish stock such as butterfish would reduce the pressure on wild herring and leave more food in the ocean for puffins, whales, tuna, and other herring predators. The search for better bait has been ongoing for many years, given the rising price of herring, but so far there are no obvious choices, and the tradition of using herring for lobster bait runs strong.

The changes in fish populations are happening so fast that, were we not watching carefully through our feeding studies, we could easily miss such events as the shift to butterfish and declining numbers of herring in puffin and tern diets. Sitting in an observation blind for three or more hours at a time takes patience and care, but the studies of fish dangling from tern and puffin beaks have now been going on for twenty-five years, and they are paying off with new discoveries each year.

Even if puffins decline at the southern frontier of their

range in Europe, this change does not necessarily mean that the effect of a warming climate will have the same outcome in Maine. Nature is full of exceptions because of complex food webs. For example, Kristen Bowser, a doctoral student of Tony Diamond's at the University of New Brunswick, created a food web for puffins breeding at Machias Seal Island. She used DNA codes to identify the prey in puffin feces. Much to her surprise, she identified evidence of fifty-one species of marine life—far more than puffins were thought to eat.[16]

This surprising food diversity does not suggest that the puffins were directly feeding on so many species. Instead, it reflects the diet of the principal prey species. Here the favored food is Atlantic herring, which feeds on plankton and the eggs and larvae of creatures ranging in size from copepods to cod. Even starfish were represented in the puffins' diet. This does not suggest that puffins eat starfish, but starfish DNA was passed through the food web to the puffins. Climate can affect any part of this complex food web and send ripples in many directions that could affect puffins.

Likewise, microclimates could have mega-effects. The greater success of puffins breeding at Eastern Egg Rock than those at Matinicus Rock and Seal Island in recent years suggests that Egg Rock puffins are feeding on different foods and perhaps in different locations than those nesting at neighboring islands, but so far the feeding places for Maine puffins remain largely a mystery.

On Eastern Egg Rock, the struggle is hardly about just the puffin. Our humble seven acres of rock is also home to one of the largest colonies in Maine of roseate terns, an endangered species in the Northeast. They are currently well established on the rock under our current management, but this population comprises half of the roseate terns in Maine. Should tragedy befall them here, it may take decades to recover, because the center of the New England roseate tern range is in coastal Massachusetts and Long Island, New York. Most of these elegant birds nest on just three islands: Tern and Ram Islands in Buzzard's Bay and Great Gull Island in Long Island Sound. A catastrophic event such as a predator or shift in food supply could have a huge effect on these colonies despite decades of focused management. Or a slow erosion of the colonies can also happen because of chronic predation or degradation of nesting habitat from invasive vegetation. That seems to be happening at Ram and Tern Islands, where the number of roseate terns has declined by more than a third from 2,124 pairs in 2000 to 1,336 pairs in 2013.[17]

Roseate terns are under assault in North America because of reduced habitat from predatory gulls, black-crowned night-herons, and invasive plants that overrun their colonies. But when they are protected, they can raise significant numbers of young—often in cleverly built artificial housing such as stone huts, wooden boxes, and even half-buried tires. It's likely their greatest problems occur during migration and on their wintering grounds.[18] But winter homes for New England's roseates have been confirmed only in recent years, such as one large congregation off the coast of Brazil that was found in the mid-1990s.[19] And here it is not clear exactly what problems they face that could prevent them from returning to their New England nesting grounds. There is much speculation, including a few reports of people shooting and eating the birds, but these are probably only a few of the many problems they encounter along their long migration route.

For roseate terns to survive to reproductive age, everything must work throughout the year at the breeding, wintering, and migration stopovers. Even ocean habitat must be viable for the birds to survive their passage. And throughout their journey, they are vulnerable to hurricanes and tropical storms. The roseate tern is symbolic of a bird caught between the over-developed North and an underdeveloped South, where poverty can lead to people repeating the very behavior that eliminated so many North American birds in centuries past.

With so many coastlines likely to be inundated by rising sea levels over the coming century, with lobsters steadily moving toward Canada, and with forage fish schooling out of the range of many birds, I have to face some obvious questions: Will the puffins one day flee Eastern Egg Rock? Despite all our work, will climate displace puffins from Egg Rock as surely as the hunters' nets scooped the last one off the island around 1880? Would my earliest skeptics who questioned the foolishness of reestablishing a northern bird on the southernmost edge of its range have the last word and say, "I told you so"?

Half a century ago Roland Clement commented in a book review of Rachel Carson's *Silent Spring* that the time had come to recognize that technology has given us such powerful tools that they are beyond the capacity of humans to use them intelligently. In a conversation with Derrick in 2011, Clement, still sharp at ninety-nine, observed that "Project Puffin was a stopgap measure to provide the possibility of things. Regardless of the eventual outcome, Project Puffin proved that there is still room for puffins in Maine—for now."

For now. Stopgap. Clement is right. The restoration of a puffin breeding colony at the edge of its range in Maine, my original goal, is no longer the real accomplishment I seek. I want puffins to inspire people to make room for wildlife—to reject the status quo that results when aggressive scavengers replace specialized, migratory species. We once believed that putting aside land for national parks and laying down arms against plume birds would restore nature. We should know better now. Humans have tilted the odds against many species to the point where there is no such thing as balance or sustainability without ongoing management.

When I narrate seabird cruises to Egg Rock, I have told the eager puffin watchers, "Every puffin we see is a miracle." It is a miracle because of the unknown obstacles and patience it took to achieve the restoration. But now I also see that the miracle is also in the wildlife stewards who keep people off plover beach nests, the state and federal biologists who wade through marshes on the hunt for invasive species, and vigilant college interns who live on isolated islands and stand guard against predators, invasive plants, ocean plastics, and disruptive human visitors.

Active restoration is the next era of bird conservation, following the passive approaches of protective laws and establishing sanctuaries. Restoration is necessary to lend a hand to the specialized, highly migratory species such as puffins that are slow to breed, are reluctant to colonize new islands, lay only one egg, and face many threats. Tony Diamond summed up this sentiment well when he said, "It's time somebody played God after our predecessors played the devil for so long." Bill Drury, who backed my early efforts to move puffins from Canada, took the God analogy a step further. He conducted some of the first active efforts for gull management in the 1960s by introducing raccoons and foxes to islands and said, "If we don't play God, the gulls will."

Centuries ago big predators such as gulls, eagles, otters, and the now-extinct sea mink pushed seabirds from one island to the next. But then the seabirds had many other islands to move to, and because they were so mobile, they could easily renest in the same year when necessary. Such movements are less feasible now because most seabird islands in Maine are dominated by big gulls and even bigger eagles who would happily snatch up a few lonely puffins and terns that were brave (or foolish) enough to attempt to start a new colony.

The tendency of rare seabirds to breed in just a few colonies makes them even more vulnerable to such inevitable catastrophes as hurricanes, typhoons, tidal waves, heat waves, and extreme rains that cause the collapse of a nesting season. The sheer number of people and the commensal scavengers and predators that shadow us have, in effect, removed the elasticity from nature's ability to bounce back following natural disaster. The birds' ability to recover following a catastrophic event is reduced further because of year-round human-imposed stress from industrial fishing, marine pollution, and climate change. These stresses can act together to reduce the birds' chances for survival at the nesting, winter, and migration stopovers.

Although I am convinced that nature holds enormous capacity for adaptation and resilience, at every season there are limits and hurdles over which wildlife must pass. When populations shrink to a fraction of their former size and ranges contract to the point that a species nests, winters, or migrates to just a few places, the survival risks increase to the point that a random event—such as a "perfect storm"—will eventually cause extinction.

Carl Buchheister, who gave me great inspiration early in the project, once said that bird restoration work was "on the side of life." When questioned what he meant, he explained restoration was good because it helped to maintain the diversity of life on earth. Today, diversity is not a mere tally of species. It is about restoring the elasticity. It is about whether we are adding, in a metaphoric sense, enough eggs to enough basketo throughout a creature's range to help it thrive. Only then can we have some real hope that a threatened species will back away from the precipice of extinction.

There are many bright signals that give cause for hope that an ethic for stewardship is emerging and will eventually prevail. I find it heartening that, after a hundred-year absence, puffins not only are back in Maine but have passed the thousand-pair mark. This success, along with the proliferation of our methods worldwide, gives me confidence that humans have the capacity to expand the habitat occupied by wildlife rather than just documenting contracting ranges, loss of nesting colonies, and eventual extinctions. This is heartening support for believing that people can develop a conservation ethic for saving life on earth.

A good place to start thinking about stewardship is to reject the idea of the "balance of nature." This enduring myth suggests that nature has purpose, that each species has a rightful place, and that, when left alone, "balance" will decide which species survive and which disappear.[20]

In contrast, a belief in active stewardship leads to rejecting the balance of nature paradigm, replacing it with the idea that each species is doing its best to survive, not because it is part of a great plan, but because it simply can. Bill Drury believed that the only constants in nature were change and chance. Everything changes, and chance events usually determine outcomes.[21]

Chance events led me to Machias Seal with Mary Majka, then to Hog Island, where I stumbled on Ralph Palmer's history of Egg Rock puffins. And it is stewardship that brought the puffins back to Egg Rock. In a world where change is the driving certainty, will stewardship prove sufficient to keep the puffins on Egg Rock? The idea of sustainability is not a steady-state balance. Not with so many humans pulling on the marine food web in so many directions. A middle-of-the-food-chain bird such as the puffin needs friends to give it time to adapt to change.

Forces beyond Eastern Egg Rock will likely someday change the coast sufficiently to displace many of the birds that nest there today. We can only guess what those forces might be or when that might happen, but in the meantime active stewardship is our best hope for keeping the puffins on Egg Rock. Humility should guide stewardship as we remind ourselves that every species is connected to dozens of other sea animals large and microscopic and that people are affecting life everywhere on earth in ways that we are only beginning to understand.

I picked the puffin for my life pursuit more than forty years ago because it captured my imagination—just as it continues to lure thousands aboard puffin-watching tours along the Maine coast. Although there are still more puffins on T-shirts and trinkets than on nesting islands, rare seabirds are coming back because we have done some of the right things with ample patience and perseverance. Saving puffins means protecting nesting habitat *and* saving their forage fish and the entire food web that supports the fish. It also means protecting oceans from pollution and rapid climate change, allowing for species to adapt to inevitable changes.

Because of their upright stance and by-chance comic manner that mimics humans, puffins can help to engage people in the conversation and actions necessary to protect oceans. As surely as my mentor Bill Drury pushed "econologists" past old notions of nature taking care of itself and advocated for direct human interventions, believing in a so-called balance of na-

they will not save the seas. Truly saving puffins also means understanding that everything is connected and especially so for
poisons, plastics, chemicals, medicine, and food waste. Everything eventually finds its way to the sea. Oceans cannot survive to nourish puffins or people without our commitment to
action. Saving puffins and other marine wildlife means that we
must strive to better understand the complexity of ocean life
and become responsible stewards.

I hope that the restored puffin colonies in Maine represent something much more important than the opportunity
of viewing this picturesque bird. I hope that Maine puffins will
help people to look at places where animals once were and
realize that we don't have to accept the diminished condition:
people *can* bring back lost species. And I hope that somewhere
out there, even Ralph Palmer is finally nodding his head in
approval. Perhaps now he might agree that Eastern Egg Rock
and its puffins represent a touchstone for people on a planet
with very wobbly legs.

14

My Skink, Christina's Ducks, and Juliet's Tern Concerto

"I never wanted to kill anything," Sue Schubel said as she recalled a day in 1988 when she was living on Eastern Egg Rock, supervising the field crew. She noticed that terns were hovering oddly, dipping down, then hovering again, but never landing. On closer inspection, she found twenty-seven terns dead on their nests. Then she saw the mink! The two-foot-long animal stared at her for a moment before slipping under a mass of boulders. After fourteen previous summers of living on the rock, we had not seen a four-legged mammal—not even a mouse. In that instant, Sue was faced with a threat that could destroy everything we had created.

She did everything she could to scare it off the island, but nothing worked. Mink are members of the weasel family, and this mink was literally a weasel in the hen house. Sue and I were committed to getting the sleek animal off the island— but how?

Days would pass without a sighting, yet she could follow the creature's movements and knew when it came out of its den under the boulders. We feared that it would not leave the island and began to try to "think like a mink." Why would it want to leave the island when it was surrounded by food? Sue tried to bait it into live traps, but to no avail—it had an abundance of seabirds within reach. It seemed that nothing was more tempting than a tern. Even baiting the traps with proprietary mink scents that hunters swear by on the mainland, including the local favorite, "Old Buck's Death," was ineffective. We eventually realized that we had to either kill that mink or risk having it kill our few precious puffins—after all our efforts, there were only sixteen puffin pairs nesting at Egg Rock in 1988. Finally, the mink met its end when it stepped in front of Sue's shotgun.

This experience was a sobering lesson in how vulnerable a restoration project such as ours is to the unexpected vagaries of chance. Twenty-six years have passed without another mink at Egg Rock, but predations are not uncommon at some of the islands closer to the mainland. For example, our tern restoration project on Jenny Island in the mouth of the New Meadows River near Cundy's Harbor has been challenged by as many as three mink in one season. This island is just two miles from the mainland and even closer to nearby forested islands that are within easy swimming distance for a mink. Yet in most years, there are no mink encounters.

Inspired by her experiences at Egg Rock, Sue Schubel has built solar-powered sound playback systems that have helped to attract twenty-seven seabird species to new nesting locations on seabird nesting islands in such varied locales as the Galápagos, China, Bermuda, and a Florida beach that was so trashed by partiers that it was nicknamed "Beer Can Island." She was in the right place at the right time to become the go-to person for building rugged audio systems that can withstand

"Seabird Sue" Schubel with one of her Murremaid Music Box audio systems installed at Eastern Egg Rock to broadcast the calls of endangered roseate terns. Photo by Stephen W. Kress

the extreme conditions where seabirds nest. More than fifty of her "Murremaid" audio playback systems are luring birds to better homes in four countries.

Sue is also the school outreach instructor for Project Puffin, partnering with Pete Salmansohn, who joined the Hog Island staff in the early 1980s and is now the project's education coordinator. Both have won national awards for youth education, and Pete has written children's books with me on puffins and bird restoration. Through their teaching, they share not just the cute and cuddly side of puffins but the practical realities of working long hours, waiting years for success, and living on remote islands where entertainment comes mostly from watching wildlife.

Sue's accomplishments in seabird management and edu-can demonstrate how inspired careers can arise from child-hood experiences—her dad, Jerry Schubel, operates aquariums, so Sue was exposed to marine life throughout her childhood. Later, her direct seabird protection experiences (such as her encounter with the Egg Rock mink) galvanized her career as biologist and educator.

For many years, Derrick and I have set aside a day to chat with the Egg Rock interns to see what led them to spend their summer on this tiny guano-splattered rock—a place that might seem a prison for most college-age students. Their stories have common threads and provide insight about the origins of an ethic for wildlife stewardship with surprising similarities to my own encounters with skinks and other creatures when I was a kid growing up in Columbus. These stories offer hope for a steady stream of future wildlife heroes who will act to save wildlife when given a chance.

My first Project Puffin assistant, Kathy Blanchard, has been working for the past two decades in seabird conservation off Quebec's north shore of the Gulf of Saint Lawrence, helping isolated fishing communities in Newfoundland and Labrador create sustainable lifestyles and find ways to coexist with the remaining fish, birds, and mammals. This was the woman who was once told by her high school guidance counselor that she would never find a job in conservation. Against the odds, she found her calling in her native Newfoundland.

When Kathy began her work there in the late 1970s, the population of puffins on islands in the region had plummeted from sixty-two thousand to fifteen thousand, because the consumption of seabirds was still important to the culture. Seabirds were linked to tradition, identity, social norms, recreation, and feelings of self-worth. When she began conducting

her research, 70 percent of families said they "harvested" seabirds and eggs. When asked how many birds were needed a
year, the average family answered: forty-four.

By 1988, after years of intensive work with local communities, the number of puffins had rebounded to thirty-five
thousand. Likewise, the numbers of common murres, razorbills, and kittiwakes all doubled in response to a sharp drop in
hunting.

Tom French, one of the architects and builders of the
puffin-rearing condos at Egg Rock, has since become a top
state biologist in Massachusetts. Now in his sixties, he says that
he "has the best job in the world" and still gets "a little bit of the
adrenaline rush when he climbs into bald eagle nest trees and
rappels off cliffs, buildings, and bridges to get to peregrine nest
sites for banding."

Kevin Bell, after leaving Project Puffin in 1975, took a
position as the youngest curator ever at Chicago's Lincoln Park
Zoo. In 1993, he became director of the entire zoo, which he
has run ever since. It was his father, Joe, who suggested back
in 1973 that we supplement the fish for our first puffin chicks
with vitamins E and B-1 and multivitamins.

Once known as the "boy in the Bronx with nearly 3,000
pets," Kevin still recalls his 1975 trip to Great Island, Newfoundland, to help collect puffin chicks. He recalls, "It was as if
I was in the middle of a beehive, except with a huge swarm of
birds. You can't even begin to concentrate on any one animal;
you're amazed at the multitude." He has since followed his passion for wildlife with his own obsession. He has dedicated
himself for three decades to saving the Bali mynah, a bird that
crashed to just twelve birds in the wild in the 1990s because of
poaching for the caged pet trade. Kevin became a global leader

in a captive breeding program to preserve the birds and release them back into the wild; he continues to be on the front lines of global efforts to stop the illegal trade in wild birds.

Richard Podolsky went from being an obsessive night netter of Leach's storm-petrels to helping seabirds in the Galápagos and Hawaii. His work with Laysan albatross at Kilauea Point was the first to demonstrate that the great birds would respond to models and sound. He then went on to become a consultant on protecting birds in the design and construction of wind turbines, power lines, and reflective glass in skyscrapers such as those in New York City's new World Trade Center complex.

Such work has never been more important. For all of their agility, birds continue to suffer from nearly every form of human enterprise. The National Audubon Society and American Bird Conservancy have calculated just some of the accidental bird deaths caused by our modern lifestyles. Collisions with glass, for example, may kill about a billion birds each year in the United States, a statistic that is second only to the deaths by feral and pet cats, which kill at least 1.4 billion birds annually in the United States.[1] Add in the continued loss of forests and wetlands to suburban development, and we are confronting a new version of *Silent Spring* that also includes summer, winter, and fall in a continual massacre of warblers, sparrows, doves, jays, cardinals, woodpeckers, and thrushes.

Reflecting on his early experiences with Project Puffin, Richard said, "There is nothing harder I will ever do than what we did on Egg Rock. Cutting and hauling mounds of sod to create our puffin nesting condos, landing in small boats in often extreme conditions. There were always risks getting in and out of boats and plenty of bumps and scrapes. In fifteen years of landing on Great Island, Newfoundland, anyone could have fallen a hundred feet down those steep slopes onto the rocks. Disaster loomed around every corner collecting and taking care of the chicks. We were creative, with almost nothing."

Richard also observed that puffin restoration was fundamentally different than the projects designed to restore peregrine, condors, and eagles. "These birds are top predators; seabirds are more complicated—much more than we appreciated in the early years of the project. That reality grew on us as we learned that we had to suppress gulls and then attract terns for the puffins. What I learned is that without management and much more involvement in rebuilding wildlife communities, specialist species such as puffins and terns are at a great disadvantage."

My greatest hope was that in some way each of my dedicated interns would find inspiration from their participation with Project Puffin to make a contribution someday to saving species. Richard Podolsky, Kathy Blanchard, Tom French, Diane DeLuca, Sue Schubel, Evie Weinstein, and many others who helped me in the early years of the project made an enormous contribution to my vision of bringing the puffin back to Maine. As their accounts clearly point out, the experience likewise made a profound impact on their thinking and actions far from Maine. That they just accepted and believed in my faith that the puffins would come back to Eastern Egg Rock leaves me deeply humbled. In June 1977, Tom French expressed this faith when he looked up to see the first puffin return to Egg Rock, not so much with a sense of wonder, but as a long overdue and expected event. Tom's response was more like, "Well, it's about time! This is what you told us would happen." Meanwhile, I was speechless with wonder that my dream had just transected with reality.

More than five hundred interns have followed these first "puffineers." Living among the seabirds, remote from the convenience of mainland life, they are rewarded daily by the flurry of wings and the screams of thriving seabird colonies that would not be present without restoration. One reality becomes clear when living amid a thriving, restored colony: a do-nothing approach to saving species usually favors generalist species such as gulls that thrive in the shadow of humanity. In contrast, specialists such as puffins and terns can do well, but only with a helping hand. Understanding why some people care more than others and encouraging a stewardship ethic will ultimately determine the diversity of life on earth.

The last of Christina Donehower's pet ducks is now dead. The final one lived until 2009, when it was at least twenty years old. When Christina was a child and teenager, the ducks followed her out onto the Puget Sound waters off Gig Harbor, across the water from Tacoma, Washington. The ducks swam alongside her kayak for a couple miles or followed her out on walks onto the mudflats. She was so attached to her ducks that she occasionally took them to school.

"Growing up, I always played this caretaking role," she said. "Throughout my childhood, I always had several domestic ducks as pets. They were dear family pets, raised from tiny ducklings so that they imprinted on me and were very tame. There were many ducks over the years, each with his/her own unique personality. Taking them down to the beach or kayaking with them were such common occurrences that they were just a regular part of the daily routine.

"I had to keep my eye on all of them, as there were foxes, raccoons, eagles, neighborhood dogs, and other predators in the area that posed a serious threat to the ducks." Christina would grow up to use that eye as one of my leading observers of gull predation on the Project Puffin islands in the early 2000s. In her research, she identified as a tern terrorist one particular herring gull, nicknamed Split Tail for some missing feathers in the middle of its tail. Throughout the whole 2004 season, this gull became a nemesis to her and Eastern Egg Rock supervisor Ellen Peterson.

"Nimble Split Tail would slip under the defending terns, enter the colony, and snatch a tern fledgling. Then a great black-black would come along and steal that tern from the herring gull," Christina said. "Then the herring gull would go back in, get another tern chick, and a great black-back would steal that one too. It kept repeating over and over, but Split Tail was very wary of people. Just when I'd get my aim [with her rifle], it flew away. Plus there were so many days it was foggy or there was boat traffic or other poor weather conditions so we couldn't risk taking shots."

Ellen said Split Tail was so elusive that it earned Christina's highest respect. "She loved that gull," Ellen said of Christina. "She kept saying how Split Tail was one of the most amazing predators she'd ever seen. It was by far the most notorious herring gull on the island. It was so agile and clever that it actually stole a lot of chicks *back* from the black-backs.

"Christina would go on the roof with her scope from four o'clock in the morning until dark. Every day had some discussion of Split Tail. Christina would come down for a break and have her Twining's English breakfast tea and I would have coffee and she would tell me how many terns Split Tail had taken. At dinner she'd tell me again about Split Tail that it did this and that. She'd say how over and over again it would creep in to a tern nesting area and all of a sudden the terns would feel its presence, but before they could react, Split Tail had grabbed a chick and was gone in an instant. All day long she saw chicks gobbled up."

Finally, on August 1, 2004, Christina recorded this in the journal: "At 15:00, I saw Split Tail, a well-known predatory HERG [herring gull] most easily recognized by a broken primary on the left wing. This bird hunted daily in the tern colony in the early morning and late afternoon/evening. His voracious appetite and infrequent attendance suggest that he was a 'commuter' from another island (or even the mainland). Perhaps he had near-fledging chicks of his own to feed. . . . He frequented two particular rocks on the NW corner of the island not used by any other gulls. I'm confident that this bird was present last year; his hunting behavior patterns are identical to those I noted for a predatory HERG in 2003.

"Split Tail was shot while he was in the process of consuming a ROST [roseate tern] fledger, a banded chick 77/C2 from productivity." (This entry meant that the team, possibly Christina herself, had banded that roseate tern chick in a special roped-off area where the interns could measure and weigh growing chicks.)

And yet that was not quite it. In 2005, there was another herring gull so amazingly similar to Split Tail that Christina and Ellen nicknamed it Spawn of Split Tail. After an early summer of witnessing its assault on the tern colonies, Ellen shot it from the rooftop of the Egg Rock Hilton on July 7. Spawn of Split Tail dropped out of sight, then washed up by the boat landing two days later. On dissection, Christina noted that the new Split Tail's stomach was loaded with tern chick bones.

"When I talk about my experiences, I always have to be a little careful," Christina said. "People love to hear the success

stories like bringing the puffin back to Egg Rock. And for people who are bombarded with doom and gloom, it's a good
thing to share some hopeful success stories."

Ellen Peterson grew up in Damariscotta, Maine. She traces
her fascination with wildlife back to a fifth-grade science project that was not your usual look-it-up-in-an-encyclopedia-
and-write-up-a-report endeavor. The project started in the
1980s when the family was on an outing and came across a
dead porcupine in the road. This is not uncommon in Maine,
but the car stopped and, at Ellen's pleading, scooped up the
crushed creature and brought it home. The animal was then
strapped to a board, covered with a large mesh screen, and left
outside. The beetles and other decomposers soon found the
carcass and did their work, eventually leaving a skeleton for
Ellen to reassemble.

This was one of the most memorable times of her childhood because she loved having such hands-on projects with
both living and dead things. The backyard decomposers didn't
finish cleaning the porcupine. She needed to be creative to
finish the job and decided to soak the remains in the family
swimming pool. This would be out of the question for most
kids, but she didn't give it a second thought. And her parents
didn't care either. Her dad's response was, "Sure, soak the bones
in the swimming pool to get them cleaner—good idea."

Years later, when attending Prescott College in Arizona,
one of her most memorable professors was Tom Fleischner, a
1975 Egg Rock intern who recognized her wildlife passion and
suggested that she reach out to Project Puffin.

Ellen said that when she first joined the project, she was
aghast at the need to kill beautiful creatures to save others. Her
experience with firearms up to that point had been primarily
a joke. When I asked during orientation for the 2001 interns if

they had ever discharged a firearm, she answered, "Yes." I asked her, "Where?" She said, "At a high school party." Interns from across the country giggled as she explained that it was a thing in small-town Maine for some girls to fire a shotgun into the sky off the porch.

But once on Egg Rock, Ellen at first not only couldn't pull the trigger, she couldn't bring herself to crush gull eggs. "I cried and had to stop and went back to the Hilton to pull myself together," Ellen said. At the end of her first season on Egg Rock, she remembered going to Hog Island to attend the annual Gulf of Maine Seabird Working Group conference, where researchers from nearly all the seabird islands from Massachusetts to Newfoundland report on the state of their seabirds.

"I was one of those people who thought that all animals deserve a spot on the planet and should have a happy ending," Ellen said. "I would hear about killing foxes for tern protection and thinking that's really harsh. But after seeing what the gulls do, I realized that was worse. . . . I ended up completely understanding it and was able to share my new feelings with rookies on the project who were feeling the same conflict."

The meaning of being a caretaker dawned on Ellen when she was an intern in 2001. While trapping puffins for banding, she trapped Y33—a previously banded puffin. That particular puffin would go on to live to thirty-five years old, becoming the oldest breeding puffin ever recorded in North America. Y33 was last observed in 2012, when she successfully reared a chick in her advanced age. But there was nothing unusual about this. After all, she had successfully raised twenty-five chicks in her twenty-nine years of breeding life at Egg Rock— always returning to the same humble rock crevice home. When Ellen trapped Y33 to replace her band in 2001, she remembers reflecting that this bird was the same age (twenty-four years

Interns "grubbing" for puffins at Eastern Egg Rock. Grubbing involves bending around granite, dangling upside down, and stretching into dark holes in the hope of pulling out a puffin chick. Photo by Derrick Z. Jackson

old) as she was at the time. She recalled, "It was just mind-boggling that they live that long and they are usually monogamous and have a family in a cozy home. These were my dreams as well!"

Juliet Lamb was another of the Egg Rock rookies troubled by the dilemma of "weeding the garden" of gulls. A brilliant but shy girl from Cape Cod, she graduated at sixteen as a presidential scholar at Nauset Regional High School and entered Harvard University with sophomore standing. She graduated from Harvard at nineteen and was also an accomplished French horn musician, playing in musicals such as *Fiddler on the Roof* and *South Pacific* and operettas such as *The Mikado*.

But she graduated not quite knowing what she wanted to do for her career.

"I went into college thinking if you wanted to work with animals, you had to work as a vet," Juliet said. "But the structural and mechanical part of veterinary medicine, the monkeying around with monkey intestines, didn't appeal to me at all.

"Halfway through college, I took field biology, entomology, and ornithology. I wanted to somehow bring these subjects together to understand greater trends. At first, I thought I would be a wildlife vet. I worked at a wildlife rehab center, but that experience came down to doing the same thing, treating one animal at a time. Yet treating only one animal at a time was to me just symptomatic of larger issues that I realized I wanted to be a part of."

There was also a world at Harvard that Juliet realized she was not comfortable with. At Harvard, the atmosphere was one where "proving yourself" to fellow students was "proving that you're busy." Everyone was proving they were the busiest. She discovered a culture of competition, and everything became a sense of being better than your neighbor without getting into depth about anything.

"Crazy as it might sound, you got the sense that it's not that cool to be passionate about things. Like, it's cool to be involved and to be busy, but it's also cool to say it doesn't really matter that much to you. I was never very good at that. I was surrounded by other people's lives, every minute competing about who had more money, who had more fun, who was having a better relationship than so and so because they feel they have to compete and keep up."

It also seemed that the other students didn't have a passion for the science. She recalled one biology class where one woman would consistently raise her hand and ask how the subject of the moment related to the next exam. In all of these classes, Juliet looked for interesting facts that would help her understand greater trends in biology.

Juliet entered another uncomfortable world in Project Puffin, arriving in 2006, when Christina became Egg Rock supervisor. Juliet had never shot a firearm before coming to us. During her orientation, when she was asked if she'd be comfortable around the shooting of predatory creatures, even if she didn't actually pull the trigger, the Cape Cod native said: "I had to really think about that. My senior thesis was why we should never have to shoot coyotes, even if they were eating things like tern eggs. I was thinking if you kill off coyotes, you have to look at all the ripple effects, and part of the problems we've created is because we're doing so many things as a society without thinking about ripple effects. We killed whales thinking, oh there's millions more whales. But there aren't."

On one of her first days on Egg Rock, Fish and Wildlife staff came out and shot 175 laughing gulls that crowded out the terns. "I know it was necessary, but I couldn't watch," Juliet said.

Soon after, she spotted a badly crippled gull. "It seemed to look at me with its eyes saying, 'Do something. Help me.' I shot it. In some ways, that was easier than letting it suffer. But it was so hard. You never get used to it. In every case you look at this gull and you see this is a beautiful bird, and I've destroyed it and I can't help but ask myself every time, why am I protecting these other birds when I can shoot this one and think nothing of it?"

But the fact that Juliet asks those life-and-death, cost-and-benefit questions, and then comes to her own answers, is likely the basis of the best kind of environmentalism that we are going to need, not just for birds, but for all the issues facing this planet. After leaving Egg Rock, Juliet became a seabird

Big smiles at Eastern Egg Rock as Juliet Lamb prepares to band a native puffin chick. The chick was extracted for banding by Juliet and fellow puffineers Lauren Scopel and Merra Howe (far right). Behind them is Steve Kress. Photo by Derrick Z. Jackson

researcher for the Royal Society for the Protection of Birds on Scotland's Orkney Islands. She observed, "No matter how uncomfortable I was, if you're sitting there in the fog and can't shoot for five days and the herring gulls are coming and eating the tern chicks like garbage, you have no choice. Yet, all of these moments of angst were worth it when the time came to search for puffin chicks."

Juliet, like Christina and Ellen before her, ultimately became the supervisor on Egg Rock, symbolizing the scores of interwoven threads of procession in Project Puffin. A half century after I started working at Audubon camps as a dishwasher and forty years after I brought the first puffin chick transplants down from Newfoundland, most of the interns "migrated" here just as I did. I cared about puffins in the 1970s because I first cared about skinks in the 1950s. Christina and Ellen were here because they first cared about ducks and porcupine skeletons. Juliet was here because she knew she wanted to be part of "greater trends."

There is also another common feature that brought them to Egg Rock. While young, they experienced meaningful experiences outdoors with animals. In contrast, many of their peers miss this opportunity—in part because of immersion in contemporary digital society and its "virtual" electronic experiences. This change in our culture came home to me one day in 2009 when Derrick and I were on the island for one of Derrick's annual puffin reports in the *Boston Globe*. Juliet was supervisor and greeted us at the Hilton with interns Yvan Satge and Liz Zinsser.

At the time, Juliet was working on her master's degree at the University of Massachusetts at Amherst. Yvan came to us from France on an international fellowship funded by Peggy and Dur Morton. Liz had just graduated from Hobart and William Smith in upstate New York. Here they had come together on an island often shrouded in invisibility by fog, pelted with downpours, and buffeted by wind. Here entertainment consisted of watching the drama of seabird family lives and simple pleasures such as watching glorious sunsets. In this unlikely setting, Juliet composed tern concertos with her French horn and a choir of boisterous terns. Derrick asked them what they remembered about their childhoods that helped them appreciate nature to the degree that they relished their isolated summer on Egg Rock.

With almost no hesitation, Juliet said: "We didn't watch TV."

Yvan said: "We didn't have TV either."

Liz said. "We had TV but no video games."

That is blasphemy in mainstream culture. In a comprehensive survey of media consumption, the Council for Research Excellence, a board founded by the Nielsen media research company, found that screen time for the average American viewing content on television, computers, movies, videos, and smartphones had grown in 2009 to 8.5 hours a day. The heaviest users were middle-aged people thirty-five to forty-nine years old.[2]

This statistic is troubling to me, because if parents are so consumed in digital space, how do they introduce their children to outdoor spaces? And many obstacles besides electronics are also preventing children from experiencing the hands-on experiences that I enjoyed as a kid. Besides digital distractions, some of these impediments include parents that are too busy or disinterested in nature, the perception that nature only exists in parks, isolation from natural habitats, the marketing of material culture to children, and the fear of letting kids explore nature on their own. Collectively, these obstacles are contributing to a growing trend of disconnection and alienation from our natural world.[3]

I certainly did not start out with this as a goal, but four decades after signing on my first research assistant, I realize that Project Puffin is not only a refuge for puffins but also an outpost for young adults who as children were able to play and explore outdoors without fear of getting muddy while putting their hands on frogs, salamanders, and other small creatures.

Back on Eastern Egg Rock, when Derrick asked Juliet, Yvan, and Liz why they were predisposed to loving nature, an outpouring of creative-play memories flowed: playing hide-and-seek in cemeteries; feeding an orphaned baby bird with scrambled eggs; picking mushrooms with grandparents; playing in tree houses; making maps; catching lizards; and chasing birds.

"When my mother got bored with us," Liz said, "she gave us a saltshaker and sent us outside, saying if we sprinkled salt on a bird's tail, we would catch it. They always flew away. But it made me always want to hold a bird."

Juliet said, "Watching wildlife is like an addiction, to see life in all that complexity. It's free, easy to see, and available to you, depending on where your imagination takes you. . . . Even with the weather, I'd rather be here than anywhere else."

More often than not, interns come to realize that life on a tiny bird island is an alternate universe that few other than researchers and a handful of curious fishers and sailors know exist. Years after she left Egg Rock, when Derrick asked Juliet if she felt she had left a certain world behind when she was on Egg Rock, she answered, "When I was on Egg Rock, I kept in touch with a few people that were off-island, but for the most part they had lives that I couldn't relate to. Most people have jobs and a life outside of their job. When you do fieldwork you're there and that's your whole existence 24/7."

Sam Radcliffe, a 2004 volunteer, remembers being enchanted as a child with the horse-and-buggy simplicity of the Amish when his parents would travel through Pennsylvania. He kept poison dart frogs as a child, much as I kept my backyard menagerie of frogs and turtles. Sam liked birds so much that his parents helped him "adopt" one of the Egg Rock puffins at the age of six, and he kept the stuffed toy puffin that came with the adoption for years. His parents sent him to Hog Island youth camp in the mid-1990s when he was twelve, where he happened to be a roommate of my son Nathan.

Years later, as an intern for Project Puffin, Sam sighted "his" long-ago adopted puffin on Egg Rock, billing with another puffin. "You could have collected my jaw up off the ground," said Sam, a graduate of Gaucher College in Baltimore. A 2011 Egg Rock intern, Jennifer Ma, grew up in Sheepshead Bay, Brooklyn, and went to Brooklyn Tech High. In her urban setting, she was fascinated at the age of six by house sparrows nesting in an awning of her home. The sparrows became so comfortable with Jennifer that they would land on her head. She would go on to be the captain of a Brooklyn Tech team that won the 2007 New York City Envirothon with a project on renewable energy in the aquatics, soils, and forestry categories. Jennifer attended the State College of New York's College of Environmental Science and Forestry in Syracuse and graduated in 2011 with a SUNY chancellor's award for student excellence.

Just over five feet tall, Jennifer endured a lot of "short" jokes in college, especially when she led visitors and prospective students through the winter snowdrifts as a student ambassador. "I'm really lucky that my boyfriend is a forest ecosystem science major and was a Boy Scout," Jen said. "But there are people that when I bring up global warming, they say, 'Oh, you're lecturing again.' That's kind of heartbreaking. A friend of mine makes a lot of money and we had a monthlong debate about climate change. It's amazing how many people are so disconnected and are so stuck in the here and now that they don't want to hear or learn. One person I'm close to said to me, 'Why do you want to spend your summer camping for minimum wage?' My answer usually goes something like this: 'Sometimes when I look at the terns and the puffins, it's so obvious they're part of a great, gigantic world and in their own

Banding tern chicks requires the donning of creative headgear to deter div-
ing birds and clothing appropriate only in a seabird colony. Left to right,
Jennifer Ma, Ivan Mota, Rolanda Steenweg, and Mike Whalen. Photo by
Derrick Z. Jackson

way, they are part of something so big—so beautiful—that I
am left in complete awe.' But my friends usually don't understand—and I know that this is their loss."

Another 2011 Egg Rock intern, Mike Whalen, took over
the family bird feeder as a child in Deep River, Connecticut,
because "my parents were always too lazy." He took what he
calls "bad" Kodak pictures of the goldfinches, blue jays, nuthatches, cardinals, chickadees, house wrens, juncos, and tufted
titmice that flocked to his seeds. He became obsessed with trying to keep varmints off the feeder. "It took me years to get my
bird feeder raccoon-proof," Mike said. "So I had to trap them
while figuring it out. We caught one and took it twenty miles
away. Not the usual thing kids were doing."

Jennifer Ma admires a tiny tern chick for a moment before banding. Photo by Derrick Z. Jackson

Rolanda Steenweg, the 2011 Egg Rock supervisor, found comfort on the seven-acre island while she gained training and experience in seabird conservation methods. A recent graduate of Dalhousie University in Halifax, Rolanda applied for a research assistant position because it would afford her the opportunity to learn special skills. These gave her a chance for leadership in ways she might not have growing up in Nova Scotia. "There's a culture where I come from where the women are just not taken seriously. I got called 'Toots' a lot by middle-aged guys."

The 2012 research team on Eastern Egg Rock was no different. Kate McNamee, a junior at Colby College in Maine, grew up in a small town at the foot of New Hampshire's Mount Monadnock, the closest three-thousand-foot mountain to Boston. She swam before she could walk and was hiking the peak by the time she was two years old. Egg Rock supervisor Maggie Post, a twenty-five-year-old graduate of Reed College in Portland, Oregon, grew up in Walnut Creek, California. She was led through the woods as a little child by her older sister. They would take long, uncut blades of grass or weeds and twist a noose at the end. When she was just six years old, she and her sister would crawl stealthily up to lizards like alligators and fence swifts. Sometimes Maggie's sister would distract the lizard as Maggie dropped the grass noose from behind over its head. "When I finally told my dad what we were doing, he said, 'Oh, I did that too with my brother when I was a boy.'"

At that point, Kristina McOmber, who had just graduated from Pomona College in Claremont, California, interjected, "That's so crazy. We did the same thing in college—catching lizards." Kristina continued, "Because my parents were always busy, I was a weird kid doing what I could on my own [in Foster City, California]. I talked to birds and built little houses in the dirt for worms to crawl in and out of. Other kids were jumping rope, playing basketball, and playing foursquare. I did some of that too, but I became known as the kid who plays with worms."

Kate laughed. "One of the things my sister and I would do was go along streams where we would set up miniature scenes with whatever material we could find. Ferns became trees. Little sticks became logs. Then we'd place a newt in it. And it would be this giant newt in our little scene."

Kate said her parents allowed her only a half-hour of television a night "and then it was back outside and we'd go find a woodsy swamp." Maggie's parents, an elementary school teacher mother and a father who owns a company that makes components for electric vehicles, allowed no TV at all for her first ten years.

Kate, Kristina, and Maggie were so clearly on a different path of life compared with so many of their other friends. When Derrick asked them to talk about their choices, Kate said, "What is so unbelievable to me is I have this job where Steve calls in for radio check and says, 'I'll be out Tuesday,' and I'll say, 'What's a Tuesday?' Days of the week don't exist at all. I have friends who are out there killing themselves and stressing out. I have a friend who is an analyst on Wall Street who says flat out, 'I'm selling my soul for ten years and then I'll go work for the Sierra Club or something.'"

Kristina spoke next: "A friend of mine told me she sold her soul as a consultant in New York City. Whenever I talk to her, she always says, 'I can't wait to get to the mountains.' Other friends say they're miserable, and we're talking people in their twenties. And here I am, on a tiny island, where the abnormal for most people is normal for us. One gorgeous sunset after another, then a storm that almost blows us off the island, then a triple rainbow to assure us that the apocalypse didn't happen. . . .

"But to be really honest, it took a while for me to get it. I didn't have my first camping experience until I was eighteen. On my first seabird research job in the Farallon Islands, I was thinking at first, 'What am I doing out here? The birds are trying to kill me, divebombing me and pooping all over me.' I was thinking, 'Have I just signed up for the Island of Death?' Then I looked out to the ocean and saw a humpback whale breach. I said, 'Okay, I get it. This is better than Disneyland.'"

Also on Egg Rock with Kristina, Kate, and Maggie was Kiah Walker. A nineteen-year-old sophomore at Williams College, she represented a milestone for Project Puffin. Kiah was the first offspring of a research assistant to join the project. She is the daughter of Diane DeLuca, who was a Project Puffin assistant in 1981, the year the first puffin pairs nested. Diane moved on from puffins to become a biologist for New Hampshire Audubon.

"To be here now and to know my mom built the puffin burrows and the petrel burrows and saw the first ones come back to nest is pretty amazing," Kiah said. The Egg Rock Hilton has photos from the decades tacked onto the wall, one of which includes Diane. "Before I came here, I read somewhere, in a journal or something, that my mother had built a rock patio around the Hilton. When I got here, we dug in the sod. The sod was this thick [she held her hands six inches apart]. But we found the patio. Walking on these stones my mother put here makes this a special place for me."

The young women and men who are ushering Egg Rock into the second decade of the twenty-first century share the past challenges of managing gulls and living in the extreme exposure of Maine's glorious seabird islands. But they also share new realities such as rising seas and extreme weather from human-induced climate change, battling invasive plants that thrive in seabird guano, and new predators that are reinserting themselves into the coastal ecosystems.

Maggie Post, the 2012 Egg Rock supervisor, had been to Egg Rock previously and thought she knew what she was signing on for, but had no way of knowing the trials that awaited her. For those accepting the Egg Rock supervisor role, there is always a special sense of responsibility for this seven-acre outpost, because it was the place where Project Puffin had its beginnings and continues to be the "showplace" where I take donors and media to let the birds tell our story.

Knowing the significance of this special island, Maggie was feeling nervous and super-responsible even before the events of that nightmare summer began. Extreme storms, scarce puffin food supplies, an ocean heat wave that brought sea surface temperatures to record highs, and the presence of new predators made dealing with gulls pale by comparison.

The first challenge of the season of 2012 was a gale-force storm that arrived on June 5, just days after Maggie and her team had opened up the Egg Rock Hilton and set up camp. Though the storm did not even merit a name, it roared onto Egg Rock with forty-mile-an-hour winds on top of an 11.2-foot, full-moon tide—the highest tide of the 2012 nesting season.

The combined seas and high tide swamped the puffin habitat on several of our islands and may have affected some of the Egg Rock puffin burrows, but it was hard for us to tell because the nests are so far below the giant boulders. We do know that the same storm washed out most of the razorbill nests on Matinicus Rock and claimed six observation blinds and towers. Big storms, extreme rainfalls, and warmer seas are predicted side effects of climate change, and it is becoming increasingly obvious how vulnerable Egg Rock and other Maine puffin colonies are to such storms.

Recall that Egg Rock sits just seventeen feet above the normal high tide. Winter storms are typically the most extreme, and it is not unusual to find chunks of granite weighing tons shifted about from one year to the next. But summer storms that sweep the sea up into the puffin-nesting habitat were unheard of even in our four-decade perspective, which is a mere whisper in the time that Egg Rock has poked above the sea. Such storms and increasingly warm sea temperatures are cause for growing concern for seabirds. This was especially apparent in 2012, when according to NOAA reports the sea surface temperature in the Gulf of Maine reached its warmest value in 150 years.[4] In 2012 some days on Egg Rock were so warm that the puffins seemed to be cooling their webbed feet in the sea rather than sitting on the hot rocks!

Ever since Sue Schubel discovered a seabird-killing mink at Egg Rock, I knew that another such attack might have a much different outcome depending on when we discovered it and our luck at interceding promptly. Yet Sue's encounter with the mink happened in 1988, and by 2012 there had not been a repeat incident. Twenty-four years passed, permitting the puffin colony to flourish, increasing to 123 pairs by 2011.

When we set up camp in late May 2012, we discovered several sets of fresh storm-petrel wings. This was odd; most predators either swallow small birds whole or leave other tell-tale clues such as plucked feathers (a sign of owl predation). Then in mid-June, Maggie spotted a large river otter on the south end of Egg Rock and it seemed very much at home, coming and going from the boulders where puffins were nesting. Apparently, it had swum the two miles from the nearest island—or farther—there was no way to know.

I had never seen an otter in Muscongus Bay, but a nearby island is named Otter Island, and otters have been seen in growing numbers at other Maine islands in recent years. When Maggie told me about the otter, I hoped it would just go away after a few days, but this four-foot-long otter seemed to have no intention of leaving. Puffins were avoiding the south end of Egg Rock, sitting in the water and crowding up elsewhere in places that they seldom frequented. When we reported the otter to the Maine Department of Inland Fisheries and Wildlife, the state biologists were adamant that it needed to be removed immediately, especially because the island is home to Maine's largest colony of roseate terns. Live-catch traps were set up and baited, but the otter showed not the slightest interest.

Maggie and her crew were trained to shoot gulls, but the otter would disappear for hours on end in the burrows. Fearing that the worst was happening during those disappearances, Hog Island facilities manager Eric Snyder came out to Egg Rock to lend assistance. Eric, twenty-seven, a geology major graduate of Vassar College in Poughkeepsie, New York, had worked at Hog Island for most of the past eight summers and had several long stints helping out on Egg Rock. A Pennsylvania farm boy and Eagle Scout, he shot the otter, a fifteen-pound female. But he noticed that her belly was flat and her teats were prominent. He wondered if there was another otter on the island.

Sure enough, there was. Maggie's team spotted a second otter winding its way among the rocks a few days later. She was determined to remove the second otter and spent several days looking for a clear shot. She finally got her chance and took a shot, but she could not find the otter—just some blood on a nearby rock. Chilled by the thought that the animal was wounded, she could only hope that her shot had been fatal and that the disturbance and possible predation would end. Days went by without another sighting. To be sure that the otter had been removed, the state wildlife department next sent a state trapper to Egg Rock. He huddled up with a blanket for two days on the lookout without a sighting. Unfortunately, several days after the trapper left, Maggie again saw the otter working its way into some of the remaining puffin burrows. Two weeks had passed since the female was shot. The otters had terrorized the island long enough.

Eric came back out again and succeeded in shooting the otter. Sadly, the otter's stomach was found to be crammed full of puffin feathers. The two otters had cleared puffins and guillemots from about one third of the nesting habitat, causing the colony to decline from 123 pairs of puffins in 2011 to 104 pairs in 2012.

There was no way of knowing how long the otters had been on the island or if they would ever have left on their own. But their presence was a sobering reminder of how eventually, even if such assaults are decades apart, there will be another mink or otter. I am convinced that without the continued presence of resident seabird stewards and talented hunters such as Eric, all of our work to bring the puffins back will find its way into the belly of one or more ravenous predators. Without vigilance and the determination to remove such predators, puffins would likely vanish from the island again.

With luck, perhaps another twenty-six years will pass before the next arrival of otters or mink. But predators will certainly return. The only question is whether we will be here to intercede on behalf of the puffins. As the project enters its fifth decade, I am increasingly convinced that we will need to stay on Egg Rock into the foreseeable future to ensure that the puffins and other seabirds are not wiped off the island by predators. Sustainability will ultimately depend on the generosity and commitment of future puffineers and generous friends of Project Puffin.

In the closing days of the 2008 summer season, Juliet Lamb wrote a poem to Egg Rock, which included the following passages:

> Many moons ago—so the stories go,
> Dr. Stephen thought, "If men
> Drove the puffins away from Muscongus Bay,
> Might we bring them back again?"

So with hat in hand, to Newfoundland
Stephen Kress did make his way,
And the puffin stack that he brought back
Made their homes on Muscongus Bay.

Soon their numbers grew, and the terns came too,
Lured by freedom from predation,
Dr. Kress' dreams were the start, it seems
Of a whole new seabird nation . . .

None of us could say in the month of May
Where the season's path might go,
Would the puffins thrive, would the terns survive?
Given time perhaps we'd know.

But "given time," what have we learned about the long-term survival of restored wildlife populations such as the Egg Rock puffins? We have certainly brought them back—but is sustainability achievable without ongoing human intervention?

Although my original plan for restoring the puffin and tern colony at Egg Rock called for us to back off someday and let the birds thrive without ongoing human intervention, it now seems that one of the key lessons we have learned at Egg Rock is that ongoing management will likely be needed to sustain the puffins and terns. This is because the restored seabird community does not live in isolation from the surrounding ecosystem that is dominated by human impacts.

The gulls that prevented puffins from naturally recolonizing Egg Rock, for example, are still so abundant that they would soon dominate the island were the interns not there to chase them off. Enduring human presence offers the only assurance that a restored puffin and tern colony has a chance for long-term survival. And this is possible only if there is ample human infrastructure in place to fund ongoing management. Stewardship for wildlife is becoming increasingly necessary considering earth's burgeoning human population. Yet fostering future wildlife stewards is far from assured in a culture where young people are too often isolated from nature.

It is no coincidence that most puffineers trace their current environmental inclinations to special interactions with wildlife as children—just as I did chasing skinks through Blacklick Woods in Columbus. The fate of the diversity of life on earth is in our hands. It is essential to find ways to let children explore and discover nature because they are the future puffin stewards and caretakers of this precious planet.

I learned stewardship at a time when most people believed in letting "nature take its course." Although there is no greater accomplishment than to save a species from extinction, a species saved today will likely be in jeopardy tomorrow without a strong culture of stewardship.

After decades of work, I'm still left with serious questions about whether humans will grasp the enormity of the stewardship necessary to preserve the earth's remaining wildlife. Project Puffin is a testimony to the enormous effort that it takes to reassemble a lost seabird colony and the even greater challenges that must be overcome to sustain it. Just as each winning marathoner must pass the baton to the next runner, so our ultimate goal should be to pass along each species on earth successfully to the next generation.

Notes

1
Chasing Skinks

1. Dempsey, "Security Easy as Child's Play."
2. Vanderbilt, "Owl Man," 72.
3. Peterson, *Birds over America*, 118.
4. Graham and Buchheister, *Audubon Ark*, 130.
5. Farrand, "Bronx County Bird Club"; Peterson, "In Memoriam: Ludlow Griscom," 600.
6. Peterson, "In Memoriam: Ludlow Griscom," 601.

2
Ghosts of the Gallery

1. Palmer, *Maine Birds*, 293.
2. Ibid., 292.
3. Hickey and Anderson, "Chlorinated Hydrocarbons," 271–273.
4. Lockley, *Puffins*, 111.
5. M. R. Audubon, *Audubon and His Journals*, 406.
6. J. J. Audubon, *Ornithological Biography*, 107.
7. Doughty, *Feather, Fashions and Bird Preservation*, 87–88.
8. Bent, *Life Histories of North American Diving Birds*, 89.
9. Harris and Wanless, *Puffin*, 58.

10. Stolzenberg, "Puffin People," 12.
11. Harris and Wanless, *Puffin,* 168.
12. Nettleship and Evans, *Atlantic Alcidae,* 133.
13. Canadian Wildlife Service, "List of Alcid Colonies in Atlantic Canada."
14. Nettleship and Evans, *Atlantic Alcidae,* 134.
15. Ibid., 66.
16. Allen, "American Naturalist," 48.
17. Graham and Buchheister, *Audubon Ark,* 8.
18. Ibid., 9–10.
19. Ibid., 12.
20. Ibid., 13.
21. Ibid., 22–23.
22. Ibid., 20, 22.
23. Ibid., 28.
24. Ibid., 33.
25. Graham, *Gulls,* 87.
26. Drury and Anderson, *Chance and Change,* 2.
27. Ernst Mayr, Foreword to Drury and Anderson, *Chance and Change,* xi; "Physiology or Medicine 1973—Press Release."
28. Drury and Anderson, *Chance and Change,* 1.
29. Ibid., 201.
30. Kalafatas, *Bird Strike,* 60.
31. Drury and Anderson, *Chance and Change,* 197.
32. Ibid., 193.
33. Ibid., 198–199.
34. Kress and Nettleship, "Re-Establishment of Atlantic Puffins."
35. Fisher, "Experiments on Homing in Laysan Albatrosses"; Serventy, "Population Ecology of the Short-Tailed Shearwater."
36. Nettleship, "Breeding Success of the Common Puffin."

3

My Judge and Drury

1. Drury and Anderson, *Chance and Change,* 193.
2. Drury, "Population Changes in New England Seabirds," 267–313, 1–15.

5
Massacre to Miracle

1. Amory, "Coyote!," 15.
2. Marzluff et al., "Lasting Recognition of Threatening People by Wild American Crows."

6
Soaked Sod and Puffin Condos

1. Chapdelaine, Laporte, and Nettleship, "Population, Productivity and DDT Contamination Trends."

7
Waiting

1. Nettleship, "Breeding Success of the Common Puffin," 259.

8
Triumph or Tragedy?

1. Kress, "Use of Decoys, Sound Recordings, and Gull Control."

9
Puffin . . . with Fish!

1. Bennett, "Rubber Bands in a Puffin Stomach," 222; Mato et al., "Plastic Pellets as a Transport Medium for Toxic Chemicals."
2. Auman et al., "Plastic Ingestion by Laysan Albatross Chicks."
3. Webster, "Deadly Tide of Plastic Waste."
4. "Strange Bedfellows."
5. Poole, "Environmental Complex, Part III."
6. Salisbury, "Audubon Shows Its Talons."
7. "Eagles and Puffins in Resurgence."

8. Livezey, "Puffins Come Home."
9. "Puffins Are Coming."
10. Kress and Nettleship, "Re-Establishment of Atlantic Puffins"; Kress, "Using Animal Behavior for Conservation."

11
Filling the Ark

1. Drury and Anderson, *Chance and Change*, 190.

12
Project Puffin Goes Global

1. Mulder and Keall, "Burrowing Seabirds and Reptiles."
2. Furness, "Occurrence of Burrow-Nesting."
3. Jones and Kress, "Review of the World's Active Seabird Restoration Projects."
4. Ibid.
5. Harris, "Biology of an Endangered Species."
6. Dunlevy et al., "Eradication of Invasive Predators on Seabird Islands."
7. Kappes and Jones, "Integrating Seabird Restoration and Mammal Eradication Programs."
8. "Petrel Head Nick Gives Chicks a Lift."
9. Madeiros, *Cahow Recovery Program: 2012–2013 Breeding Season Report*.
10. Carlile, Priddel, and Madeiros, "Establishment of a New, Secure Colony."
11. Madeiros, *Cahow Recovery Program: 2010–2011 Breeding Season Report*.
12. Derrick Z. Jackson, telephone interview with David Wingate, 2011.
13. Hasegawa, "Conservation of Large Seabird."
14. Hasegawa, *Yomiuri Shimbun*.
15. Deguchi et al., "Translocation and Hand-Rearing Techniques."
16. Ibid.
17. Carter et al., "1986 Apex Houston Oil Spill."
18. Parker et al., "Assessment of Social Attraction Techniques."
19. Roby et al., "Effects of Colony Relocation."

20. Roby, Collis, and Lyons, "Conservation and Management for Fish-Eating Birds."

21. Roby et al., "Effects of Colony Relocation."

22. Collis et al., "Barges as Temporary Breeding Sites."

23. Chen et al., "Breeding Biology of Chinese Crested Terns."

24. Fan et al., "Restoration of Chinese Crested Terns."

25. Gummer, "Chick Translocation," 9; North American Bird Conservation Initiative, "State of the Birds 2010 Report," 13.

26. Chapdelaine, Laporte, and Nettleship, "Population, Productivity and DDT Contamination Trends."

27. Sawyer and Fogle, "Establishment of a New Breeding Colony."

28. Corrigan and Diamond, "Northern Gannet, *Morus bassanus,* Nesting."

29. Ahlering et al., "Research Needs and Recommendations."

13
Reconsidering the Balance of Nature

1. Donehower et al., "Effects of Gull Predation and Predator Control," 29–39.

2. Small and Nicholls, "Global Analysis of Human Settlement."

3. Crawford and Jahncke, "Comparison of Trends in Abundance," 145–156.

4. Jackson, "Penguin Species Faces Extinction."

5. Pikitch et al., "Little Fish, Big Impact."

6. Alder et al., "Forage Fish."

7. Jackson, "Prince Herring."

8. Cury et al., "Global Seabird Response to Forage Fish Depletion," 1703.

9. Breton and Diamond, "Annual Survival of Adult Atlantic Puffins."

10. Jacobson et al., "Maine's Climate Future," 20.

11. Nye et al., "Changing Spatial Distribution of Fish Stocks."

12. Ibid., 113.

13. Durant, Anker-Nilssen, and Stenseth, "Trophic Interactions under Climate Fluctuations."

14. Mike Harris, personal communication with author.

15. Sewall et al., "Bringing Back the Fish."

16. Bowser, Diamond, and Addison, "From Puffins to Plankton."

17. Gulf of Maine Seabird Working Group, "Minutes of the Gulf of Maine Seabird Working Group."
18. Nisbet, "Migration and Winter Quarters."
19. Hayes et al., "Nonbreeding Concentration of Roseate and Common Terns."
20. Kricher, *Balance of Nature.*
21. Drury and Anderson, *Chance and Change.*

14

My Skink, Christina's Ducks, and Juliet's Tern Concerto

1. Clem, "Collisions with Bird and Windows"; Loss, Will, and Marra, "Impact of Free-Ranging Domestic Cats."
2. Nielsen Company, "A2/M2 Three Screen Report."
3. Kellert, *Birthright*; Louv, *Last Child in the Woods.*
4. Romm, "NOAA: In 2012, Waters off Northeast U.S. Coast."

Bibliography

Ahlering, Marissa A., et al. "Research Needs and Recommendations for the Use of Conspecific Attraction Methods in the Conservation of Migratory Songbirds." *Condor* 112 (2010): 252–264.

Alder, Jacqueline, et al. "Forage Fish: From Ecosystems to Markets." *Annual Review of Environmental Resources* 33 (2008): 153–166.

Allen, J. A. "The American Naturalist, an Illustrated History of Natural History." *Journal of the American Society of Naturalists* 10 (January 1876): 48.

Amory, Cleveland. "Coyote!" Field Museum of Natural History Bulletin (January 1974): 3–7, 15.

Audubon, J. J. *Ornithological Biography; or, an Account of the Habits of the Birds of the United States of America.* Edinburgh: Adam and Charles Black, 1835.

Audubon, Maria R. *Audubon and His Journals,* vol. 2. New York: Charles Scribner's Sons, 1987.

Auman, Heidi J., et al. "Plastic Ingestion by Laysan Albatross Chicks on Sand Island, Midway Atoll, in 1994 and 1995." In *Albatross Biology and Conservation,* ed. Graham Robinson and Rosemary Gales. Chipping Norton: Surrey Beatty and Sons, 1997.

Bennett, G. R. "Rubber Bands in a Puffin's Stomach." *British Birds* 53 (1960): 222.

Bent, A. C. *Life Histories of North American Diving Birds.* United States National Museum Bulletin 107. Washington, DC: Smithsonian Institution, 1919.

Bowser, A. Kirsten, Antony W. Diamond, and Jason A. Addison. "From Puffins to Plankton: A DNA-Based Analysis of a Seabird Food Chain in the Northern Gulf of Maine." *PLoS ONE* 8, no. 12 (2013): e83152. doi:10.1371/journal.pone.0083152.

Breton, André R., and Antony W. Diamond. "Annual Survival of Adult Atlantic Puffins *Fratercula arctica* Is Positively Correlated with Herring *Clupea harengus* Availability." *Ibis* 156 (2014): 35–47.

Canadian Wildlife Service. "List of Alcid Colonies in Atlantic Canada." Excel spreadsheet by S. Wilhelm, Mt. Pearl, NFLD, 2014.

Carlile, Nicholas, David Priddel, and Jeremy Madeiros. "Establishment of a New, Secure Colony of Endangered Bermuda Petrel *Pterodroma cahow* by Translocation of Near Fledged Nestlings." *Bird Conservation International* 22 (2012): 46–58.

Carter, H. R., et al. "The 1986 Apex Houston Oil Spill in Central California: Seabird Injury Assessments and Litigation Process." *Marine Ornithology* 31 (2003): 9–19.

Chapdelaine, G., P. Laporte, and D. N. Nettleship. "Population, Productivity and DDT Contamination Trends of Northern Gannets (*Sula bassanus*) at Bonaventure Island, Quebec, 1967–1984." *Canadian Journal of Zoology* 65 (1987): 2922–2926.

Chen, Shui-Hua, et al. "The Breeding Biology of Chinese Crested Terns in Mixed Species Colonies in Eastern China." *Bird Conservation International* 21 (2011): 266–273.

Chen, Shui-Hua, et al. "A Small Population and Severe Threats: Status of the Critically Endangered Chinese Crested Tern." *Oryx* 43 (2009): 209–212.

Clem, Daniel, Jr. "Collisions with Bird and Windows: Mortality and Prevention." *Journal of Field Ornithology* 6 (1990): 120–128.

Collis, Ken, et al. "Barges as Temporary Breeding Sites for Caspian Terns: Assessing Potential Sites for Colony Restoration." *Wildlife Society Bulletin* 30 (2002): 1140–1149.

Corrigan, Sean, and Antony W. Diamond. "Northern Gannet, *Morus bassanus*, Nesting on Whitehorse Island, New Brunswick." *Canadian Field-Naturalist* 115 (2001): 176–177.

Crawford, R. J. M., and J. Jahncke. "Comparison of Trends in Abundance of Guano-Producing Seabirds in Peru and Southern Africa." *South African Journal of Marine Science* 21 (1999): 145–156.

Crawford, R. J. M., et al. "Collapse of South Africa's Penguins in the Early 21st Century." *African Journal of Marine Science* 33 (2011): 139–156.

Croxall, John P., et al. "Seabird Conservation Status, Threats and Priority Actions: A Global Assessment." *Bird Conservation International* 22 (2012): 1–34.

Cury, Philippe M., et al. "Global Seabird Response to Forage Fish Depletion— One-Third for the Birds." *Science* 334 (2011): 1703–1706.

Deguchi, Tomohiro, et al. "Translocation and Hand-Rearing Techniques for Establishing a Colony of Threatened Albatross." *Bird Conservation International* 22 (2012): 66–81.

Dempsey, Hood. "Security Easy as Child's Play." *Columbus Dispatch,* July 25, 2002.

Donehower, Christina E., et al. "Effects of Gull Predation and Predator Control on Tern Nesting Success at Eastern Egg Rock, Maine." *Waterbirds: The International Journal of Waterbird Biology* 30 (2007): 29–39.

Doughty, Robin W. *Feather, Fashions and Bird Preservation: A Study in Nature Protection.* Berkeley: University of California Press, 1975.

Drury, William H. "Population Changes in New England Seabirds." *Bird-Banding* 44 (1973): 267–313, and 45 (1974): 1–15.

Drury, William Holland, Jr., and John G. T. Anderson, eds. *Chance and Change: Ecology for Conservationists.* Berkeley: University of California Press, 1998.

Dunlevy, P., et al. "Eradication of Invasive Predators on Seabird Islands." In *Seabird Islands: Ecology, Invasion and Restoration,* ed. C. P. H. Mulder et al. Oxford University Press, 2011.

Durant, J. M. T. Anker-Nilssen, and N. C. Stenseth. "Trophic Interactions under Climate Fluctuations: The Atlantic Puffin as an Example." *Proceedings of the Royal Society of London B: Biological Sciences* 270 (2003): 1461–1466.

"Eagles and Puffins in Resurgence." *New York Times,* January 4, 1981.

Ellis, Julie C., Jose Miguel Fariña, and Jon D. Witman. "Nutrient Transfer from Sea to Land: The Case of Gulls and Cormorants in the Gulf of Maine." *Journal of Animal Ecology* 75 (2006): 565–574.

Fan, Z. Y., et al. "Restoration of the Chinese Crested Terns in Jiushan Islands." In *Proceedings of the 12th China Ornithological Conference.* Hangzhou, 2013.

Farrand, John, Jr. "The Bronx County Bird Club: Memories of Ten Boys and an Era That Shaped American Birding." *American Birds* 45 (1991): 372–381.

Fisher, Harvey I. "Experiments on Homing in Laysan Albatrosses, *Diomedea immutabilis*." *Condor* 73 (1971): 389–400.

Fowlie, Martin. "Red List for Birds 2013: Number of Critically Endangered Birds Hits New High." Press release, BirdLife International, November 26, 2013. http://www.birdlife.org/worldwide/news/red-list-birds-2013-number-critically-endangered-birds-hits-new-high.

Furness, R. W. "The Occurrence of Burrow-Nesting among Birds and Its Influence on Soil Fertility and Stability." In *The Environmental Impact of Burrowing Animals and Animal Burrows: The Proceedings of a Symposium Held at the Zoological Society of London on 3rd and 4th May 1990*, ed. Peter S. Meadows and Azra Meadows, 53–65. Oxford: Published for the Zoological Society of London by Clarendon Press, 1991.

Graham, Frank, Jr. *Gulls: A Social History*. New York: Random House, 1975.

Graham, Frank, Jr., with Carl W. Buchheister. *The Audubon Ark: A History of the National Audubon Society*. University of Texas Press, 1990.

Gulf of Maine Seabird Working Group (GOMSWG). "Minutes of the Gulf of Maine Seabird Working Group, Annual Meeting—August 2013." http://gomswg.org/pdf_files/GOMSWG%208-12-2013%20minutes.pdf.

Gummer, Helen. "Chick Translocation as a Method of Establishing New Surface-Nesting Seabird Colonies: A Review." DOC Science Internal Series. Wellington: New Zealand Department of Conservation, 2003. http://www.doc.govt.nz/documents/science-and-technical/dsis150.pdf.

Harris, Michael P. "The Biology of an Endangered Species, the Dark-Rumped Petrel (*Pterodroma phaeopygia*), in the Galápagos Islands." *Condor* 72 (1970): 76–84.

Harris, Mike P., and Sarah Wanless. *The Puffin*. New Haven: Yale University Press, 2012.

Hasegawa, Hiroshi. "Conservation of Large Seabird, Short-Tailed Albatross." In *Conservation Ornithology*, ed. Akira Yamagishi, 89–104. Kyoto: Kyoto University Press, 2007.

Hasegawa, Hiroshi. *Yomiuri Shimbun*, August 5, 1993.

Hickey, J. J., and H. W. Anderson. "Chlorinated Hydrocarbons and Eggshell Changes in Raptorial and Fish-Eating Birds." *Science* 162 (1968): 271–273.

IUCN Red List of Threatened Species. Version 2013.2. www.iucnredlist.org.

Jackson, Derrick Z. "A Penguin Species Faces Extinction." *Boston Globe*, September 7, 2010.

Jackson, Derrick Z. "Prince Herring: Glamourizing the Forage Fish." *Boston Globe*, April 14, 2012.

Jacobson, G. L., et al., eds. *Maine's Climate Future: An Initial Assessment.* Orono: University of Maine, 2009.

Jiang, Hangdong, Lin Chen, and Fengqi He. "Preliminary Assessment on the Current Knowledge of the Chinese Crested Tern (*Sterna bernsteini*)." *Chinese Birds* 1 (2010): 163–166.

Jones, Holly P., and Stephen W. Kress. "A Review of the World's Active Seabird Restoration Projects." *Journal of Wildlife Management* 76 (2012): 2–9.

Kalafatas, Michael N. *Bird Strike: The Crash of the Boston Electra.* Lebanon, NH: Brandeis University Press, 2010.

Kappes, Peter J., and Holly P. Jones. "Integrating Seabird Restoration and Mammal Eradication Programs on Islands to Maximize Conservation Gains." *Biodiversity and Conservation* 23 (2014): 503–509.

Kellert, Steven. *Birthright: People and Nature in the Modern World.* New Haven: Yale University Press, 2012.

Kress, Stephen W. "The Use of Decoys, Sound Recordings, and Gull Control, for Re-Establishing a Tern Colony in Maine." *Colonial Waterbirds* 6 (1983): 185–196.

Kress, Stephen W. "Using Animal Behavior for Conservation: Case Studies in Seabird Restoration from the Maine Coast, USA." *Journal of the Yamashina Institute for Ornithology* 29 (1997): 1–26.

Kress, Stephen W., and David N. Nettleship. "Re-Establishment of Atlantic Puffins (*Fratercula arctica*) at a Former Breeding Site in the Gulf of Maine." *Journal of Field Ornithology* 59 (1988): 161–170.

Kricher, John. *The Balance of Nature: Ecology's Enduring Myth.* Princeton, NJ: Princeton University Press, 2009.

Livezey, Emilie Tavel. "Puffins Come Home." *Christian Science Monitor*, August 7, 1980.

Lockley, R. M. *Puffins*. London: J. M. Dent and Sons, 1953.

Loss, Scott R., Tom Will, and Peter P. Marra. "The Impact of Free-Ranging Domestic Cats on Wildlife of the United States." *Nature Communications* 4 (2013). doi:10.1038/ncomms2380.

Louv, Richard. *Last Child in the Woods: Saving Our Children from Nature-Deficit Disorder*. New York: Workman, 2005.

Madeiros, Jeremy. *Cahow Recovery Program: 2010–2011 Breeding Season Report*. Hamilton: Bermuda Department of Conservation, 2011.

Madeiros, Jeremy. *Cahow Recovery Program: 2012–2013 Breeding Season Report*. Hamilton: Bermuda Department of Conservation, 2013.

Marzluff, John M., et al. "Lasting Recognition of Threatening People by Wild American Crows." *Animal Behaviour* 79 (2010): 699–707.

Mato, Yukie, et al. "Plastic Resin Pellets as a Transport Medium for Toxic Chemicals in the Marine Environment." *Environmental Science and Technology* 35 (2001): 318–324.

Mulder, Christa P. H., and Susan N. Keall. "Burrowing Seabirds and Reptiles: Impacts on Seeds, Seedlings and Soils in an Island Forest in New Zealand." *Oecologia* 127 (2001): 350–360.

Nettleship, David N. "Breeding Success of the Common Puffin (*Fratercula arctica* L.) on Different Habitats at Great Island, Newfoundland." *Ecological Monographs* 42 (2011): 239–268.

Nettleship, David N., and P. G. H. Evans. "Distribution and Status of the Atlantic Alcidae." In *The Atlantic Alcidae: The Evolution, Distribution and Biology of the Auks Inhabiting the Atlantic Ocean and Adjacent Water Areas*, ed. David N. Nettleship and Tim R. Birkhead. Orlando, FL: Academic Press, 1985.

Nielsen Company. "A2/M2 Three Screen Report: Television, Internet and Mobile Usage in the US." Vol. 5: 1st Quarter 2009. http://www.nielsenwire.com.

North American Bird Conservation Initiative, US Committee. "The State of the Birds 2010 Report on Climate Change, United States of America." Washington, DC: US Department of the Interior, 2010.

Nye, Janet A., et al. "Changing Spatial Distribution of Fish Stocks in Relation to Climate and Population Size on the Northeast United States Continental Shelf." *Marine Ecology Progress Series* 393 (2009): 111–129.

Palmer, Ralph S., ed. *Handbook of North American Birds.* Volume 1: *Loons through Flamingos.* New Haven: Yale University Press, 1962.

Palmer, R. S. *Maine Birds. Bulletin of the Museum of Comparative Zoology* 102 (1949).

Parker, Michael W., et al. "Assessment of Social Attraction Techniques Used to Restore a Common Murre Colony in Central California." *Waterbirds: The International Journal of Waterbird Biology* 30 (2007): 17–28.

Peterson, Roger Tory. *Birds over America.* Rev. ed. New York: Dodd, Mead, 1983.

Peterson, Roger T. "In Memoriam: Ludlow Griscom." *Auk* 82 (1965): 598–605.

"Petrel Head Nick Gives Chicks a Lift." *Illawarra Mercury,* July 7, 2004.

"Physiology or Medicine 1973—Press Release." Nobelprize.org. Nobel Media AB 2014. http://www.nobelprize.org/nobel_prizes/medicine/laureates/1973/press.html.

Pikitch, Ellen K., et al. "Little Fish, Big Impact: Managing a Crucial Link in Ocean Food Webs." Lenfest Forage Fish Task Force, 2012.

Poole, William. "The Environmental Complex, Part III." Heritage Foundation Reports, Institutional Analysis #19, 1982.

"The Puffins Are Coming." *Newsweek,* August 24, 1981.

Roby, D. D., K. Collis, and D. E. Lyons. "Conservation and Management for Fish-Eating Birds and Endangered Salmon." USDA Forest Service Gen. Tech. Rep. PSW-GTR-191, 2005.

Roby, Daniel D., et al. "Effects of Colony Relocation on Diet and Productivity of Caspian Terns." *Journal of Wildlife Management* 66 (2002): 662–673.

Romm, Joe. "NOAA: In 2012, Waters off Northeast U.S. Coast Were Warmest in 150 Years." Climate Progress, April 28, 2013. http://thinkprogress.org/climate/2013/04/28/1931351/noaa-in-2012-waters-off-northeast-us-coast-were-highest-measured-in-150-years/.

Salisbury, David F. "Audubon Shows Its Talons to Representatives of Agriculture and Interior." *Christian Science Monitor,* July 8, 1981, 3.

Sawyer, Steve L., and Sally R. Fogle. "Establishment of a New Breeding Colony of Australasian Gannets (*Morus serrator*) at Young Nick's Head Peninsula." *Notornis* 60 (2013): 180–182.

Serventy, D. L. "Aspects of the Population Ecology of the Short-Tailed
    Shearwater (*Puffinus tenuirostris*)." *Proceedings of the 14th Inter-
    national Ornithological Congress* (1967): 165–190.
Sewall, Brad, et al. *Bringing Back the Fish: An Evaluation of U.S. Fisheries
    Rebuilding under the Magnuson-Stevens Fishery Conservation and
    Management Act.* NRDC Report R:13-01-A, February 2013.
Small, Christopher, and Robert J. Nicholls. "A Global Analysis of Human
    Settlement in Coastal Zones." *Journal of Coastal Research* 19 (2003):
    584–599.
Stolzenberg, W. "The Puffin People." *Sea Frontiers* 41, no. 2 (1995): 12.
"Strange Bedfellows." *Sports Illustrated*, April 7, 1980.
Suzuki, Yasuko. "Piscivorous Colonial Waterbirds in the Columbia River
    Estuary: Demography, Dietary Contaminants, and Management."
    PhD diss., Oregon State University, Corvallis, 2012.
Twain, Mark, and Charles Dudley Warner. *The Gilded Age: A Tale of
    To-Day.* Hartford, CT: American Publishing, 1873.
Vanderbilt, Sanderson. "Owl Man." *New Yorker*, November 28, 1936, 72.
Wang, Zuoyue. *In Sputnik's Shadow: The President's Scientific Advisory
    Committee and Cold War America.* New Brunswick, NJ: Rutgers
    University Press, 2008.
Webster, Bayard. "Deadly Tide of Plastic Waste Threatens World's Oceans
    and Aquatic Life." *New York Times*, December 25, 1984.
Worm, Boris, et al. "Impacts of Biodiversity Loss on Ocean Ecosystem
    Services." *Science* 314 (2006): 787–790.

Index

Accidental bird deaths, 301
African penguins, 278
Agricultural practices, changes in, 278
Agriculture Department (USDA), 99, 119, 132
Ayres, Chris, 90, 93
Albatross, 35, 53, 66, 116, 189, 228, 232, 234, 246–250, 301
Albin, Mac and Steve, 8–12, 15
Alcidae family, 39, 46, 61, *61*, 83
Allan D. Cruickshank Wildlife Sanctuary, Eastern Egg Rock named as, 130
Allen, Joel Asaph, 39–40
Allen, Rod, 177
American Bird Conservancy, 301
American Ornithologists' Union, 40, 42
Anderson, John, 222
Anticosti Island, 265
Antioch College: Glen Helen Outdoor Education Center, 25–26, 31; Glen Helen Raptor Center, 25
Apex Houston (oil barge) spill (1986), 250–251
Aquariums, 10–11
Arctic terns. *See* Terns
Arevalo, Washington "Wacho," 238
Atlanta Bird Club, 117
Audio playbacks: attracting various seabird species to new nesting locations, 297–298; cahows, 245; Galápagos petrels, 233–234, 236–237; gannets, 264; storm-petrels, 134, 172; terns, 158–160, 183, 185, 217. *See also* Social attraction
Audubon, John James, 35, 38
Audubon Christmas Bird Count, 14–15
Audubon Magazine, 19
Audubon Seabird Restoration Program, 223